UNIVERSITY OF
WOLVERHAMPTON

Harrison Learning ~~~~~
City
University **Techniques**

D1331291

Basic Bioscience Laboratory Techniques

A Pocket Guide

Philip Bonner and Alan Hargreaves

WILEY-BLACKWELL

A John Wiley & Sons, Ltd., Publication

Wiley-Blackwell is an imprint of John Wiley & Sons, formed by the merger of Wiley's global
Scientific, Technical and Medical business with Blackwell Publishing.

Registered office: John Wiley & Sons, Ltd, The Atrium, Southern Gate, Chichester, West Sussex,
PO19 8SQ, UK

Editorial offices:
9600 Garsington Road, Oxford, OX4 2DQ, UK
The Atrium, Southern Gate, Chichester, West Sussex, PO19 8SQ, UK
111 River Street, Hoboken, NJ 07030-5774, USA

For details of our global editorial offices, for customer services and for information about how to
apply for permission to reuse the copyright material in this book please see our website at
www.wiley.com/wiley-blackwell.

Library of Congress Cataloguing-in-Publication Data

Bonner, Philip L. R.
  Basic bioscience laboratory techniques : a pocket guide / Philip Bonner and Alan Hargreaves.
    p. ; cm.
  Includes bibliographical references and index.
  ISBN 978-0-470-74309-6 (pbk.)
  1. Biology—Laboratory manuals.   2. Biology–Handbooks, manuals, etc.   I. Hargreaves, Alan.
II. Title.
  [DNLM: 1. Laboratory Techniques and Procedures–Handbooks. 2. Biotechnology–Handbooks.
3. Research–Handbooks. QY 39]
  QH317.B66 2010   2010033321
  570.78–dc22

A catalogue record for this book is available from the British Library.

This book is published in the following electronic format: ePDF 9780470972052
Set in 8.75/10.5 Times by Thomson Digital, Noida, India

First Impression   2011

CONTENTS

PREFACE

In recent years there have been many advances made in the field of bioscience research. This progress has been driven by parallel advances in methodology. New and minor modifications to existing techniques can propel and enhance scientific investigation. However, the use of any procedure can only be advantageous if there is an appreciation of the basic principles, requirements, benefits and pitfalls of the techniques employed. This is true for both experienced as well as novice investigators.

We have been teaching in the HE sector for the past 20 years. During this period we have become increasingly aware that newly enrolled students are struggling to appreciate the basics of laboratory techniques and data analysis. Although the reasons for this are many fold, a sound platform for scientific investigation is still dependent upon the understanding and correct use of basic bioscience laboratory techniques. Our students perceive that there is a lack of appropriate support in the literature. With this in mind, the rationale for this book is to supply you (students) with an easy to use reference book that will provide you with a basis for understanding the basic bioscience techniques that you will come across in your HE studies.

The book provides the foundation of many basic laboratory techniques, worked examples of common calculations and (we hope) helpful tips. We wanted the book to be used prior to, during and after laboratory classes, which is why it is designed to fit into the pocket of a laboratory coat. The content reflects the size constraints and accompanying material will be available on a companion web page. In addition, supplementary reading will be needed to provide more detailed information on the techniques covered. However, we believe that the book will be of benefit to students starting their degree courses in any biosciences theme. The techniques covered in this book will be encountered by you repeatedly during laboratory sessions and we hope that the book will remain a valuable reference text throughout your degree course.

We would both like to thank our families (Alan: Olivia, Alan, Elena and Stephanie. Phil: Elizabeth, Francesca, Mark and Neil) and friends (Chris, David, John, Krishan, Lee, Mel and Wayne) for their support and suggestions during the preparation of this book. We would also like to thank Nicky and Fiona at Wiley and Sons for their patience and encouragement during the book's preparation period.

Philip L.R. Bonner
Alan J. Hargreaves

GLOSSARY

Agarose: a polysaccharide with many hydroxyl groups. The material can be used to construct gels to separate nucleic acid fragments.

Ampholytes: mixture of polycarboxylic and polyamino acids.

Bacilli: Rod shaped baceria (singular - bacillus). An example is *Bacillus megaterium.*

bis: occurring twice, as in bis acrylamide or bis Tris propane

Centrioles: Cylindrical structures that nucleate the formation of microtubules in eukaryotic cells and form the poles of the mitotic spindle.

Chelating agent: A compound (e.g. EDTA) which preferentially binds to metal ions. This reduces or effectively eliminates the metal ions presence in a solution.

Cocci: Round or spherical shaped bacteria (singular - coccus). An example is *Micrococcus luteus.*

Collminated: to make parallel

Confidence interval: A confidence interval for a set of values describes the likely range within which a specific value can be considered to be part of the same population or data set. Increasing the degree of confidence with which this assumption can be made (e.g. from 95 to 99%) will increase the limits of the range and *vice versa*. This measure is often used in clinical research, as it describes the likely range of expected values for a particular treatment or condition. See also 'confidence limits'

Confidence limits: A statistical term for the pair of values that describe the upper and lower end of a range of values within which a value can be expected to fall for a given level of confidence. In other words the two values are the end points of the confidence interval.

Dialysis: a means to separate components in a solution by unequal diffusion through a semi permeable membrane

Dynodes: one of a series of electrodes in a photomultiplier

Endoplasmic reticulum: A specialized membranous organelle within eukaryotic cells responsible for synthesis of membrane proteins and lipids.

Eukaryotes: Organisms that are composed of one or more cells which contain a membrane enclosed nucleus and organelles (e.g. animal, plant, fungi, yeast and most algae).

Focal plane: The plane perpendicular to the lens' optical axis in which images of points in the object field of the lens are in focus.

Glycogen: A storage form of glucose in animal cells. Often present as cytoplasmic granules, which are particularly abundant in liver cells.

Golgi apparatus: A specialised organelle within the cytoplasm of eukaryotic cells that is involved in the glycosylation of membrane proteins and secretory proteins.

Gravity: The natural weak force of attraction that exists between all bodies in the universe. ($g = 981$ cm s^{-2})

Homogenate: a tissue or cellular mixture resulting from the action of a homogeniser.

Homogeniser: a laboratory machine that disrupts tissue (cells) by shearing, cutting and blending the starting material in a buffer

Hydrophilic: ("water loving") a molecule (functional group) which prefer to interact with water or other polar solvents.

Hydrophobic: ("water hating") a molecule (functional group) which avoids contact with water or other polar solvents.

Immiscible: incapable of mixing together e.g. oil and water

Intermediate filaments: Filaments of 10 nm width made of various proteins (e.g desmin in muscle cells, nuclear lamins in the nuclear lamina). These filaments play a structural role in eukaryotic cells.

In vitro: Literally means "in glass"

In vivo: Literally means "in life"

Isoelectric point (pI): the pH value at which a molecule carries no net charge. The molecule will not move under the influence of an electrical field or bind to ion exchange resins.

I.U.: International units of enzyme activity defined as 1 mol of product formed (or substrate consumed) min^{-1} at a given temperature (usually 25 °C).

Lysosomes: Single membrane bounded organelles in eukaryotic cells that contain hydrolytic enzymes. These enzymes (e.g. acid phosphates) are maintained in an acidic pH environment within the organelle, where they can degrade macromolecules in a controlled manner.

Median: The exact mid point of a set of values or observations.

Mode: The most frequently observed value in a set of data.

Meniscus: the curved shape assumed by a liquid in a cylindrical tube

Microfilaments: Polymers of 5 nm width made from the protein actin. Important in contractile processes (e.g. thin filaments in muscle cells and the formation of the contractile ring in the late stages of mitosis).

Microtubules: Tubular Polymers (external diameter = 25 nm) of the protein tubulin. Microtubules play an important role in intracellular transport, cell movements and cell division, forming interphase networks and the mitotic spindle.

Mitochondria: discrete organelles present in most eukaryotic cells involved in the production of ATP.

Mycoplasma: A very small bacterium that lacks a cell wall. Mycoplasmas are typical contaminants of eukaryotic cell cultures.

Nomogram: a two-dimensional graphical calculating plot.

Non-parametric test: Statistical tests that do not make the assumption of a normal distribution of values and therefore can be applied to asymmetric distributions more robustly. Values are usually ranked or grouped. An example is the Mann Whitney U test.

Normal (Gaussian) distribution: Asymmetrical or bell shaped distribution of data values with the arithmetic mean in the exact centre.

Nucleic acids: Nucleic acids are unbranched polymers of nucleotides.

Nucleoside: The combination of a base and a pentose sugar with no phosphate groups attached. Examples (with their corresponding base in parentheses) are

- adenosine or deoxyadenosine (adenine)
- guanosine or deoxyguanosine (guanine)
- cytidine or deoxycytidine (cytosine)
- uridine (uracil) or deoxythymidine (thymine)
- Phosphorylation of nucleosides at the 5′ carbon atom of the pentose sugar results in the formation of the corresponding **nucleotides**. Nucleoside triphosphates, nucleoside diphosphates and nucleoside monophosphates are nucleotides with three, two or one phosphate groups. Examples are adenosine monophosphate (AMP), adenosine diphosphate (ADP) and adenosine triphosphate (ATP)

Nucleotide: Nucleotides are made up of three components:

- A pentose sugar (ribose in RNA and deoxyribose in DNA);
- a nitrogen containing ring structure or *base*
  In DNA the nucleotide bases are adenine (A), thymine (T), guanine (G) and cytosine (C).
  In RNA the nucleotide bases are A, T, G and U (uracil, which replaces cytosine)
  The combination of a base and a pentose sugar is called a **nucleoside**.
- Up to 3 phosphate groups.

  These are attached at the 5' carbon atom of the pentose. Both DNA and RNA are assembled from nucleoside triphosphates (i.e. **nucleotides** with 3 phosphate groups). Adjacent nucleotides in the nucleic acid sequence are covalently linked between the 3' carbon atom of one pentose and the phosphate group attached to the 5' carbon of the next pentose. The two terminal phosphates groups are released from each nucleotide as it is added on to the polymerising polynucleotide chain.

Nucleus: a eukaryotic membrane-bound organelle which contains the majority of the cells genetic material.

Null hypothesis: In statistical testing, the null hypothesis suggests that there is no statistical significance in a set of observations. It proposes that there is no real difference between 2 variables or one variable and zero that could not have arisen by chance. For the null hypothesis to be rejected, an appropriate statistical test must be performed producing a level of confidence (typically $p<0.05$) indicating that the differences are significant.

Oil immersion: This is a technique that involves placing a drop of immersion oil, which is transparent and has a high refractive index, between the $\times 100$ objective lens and the cover slip covering the specimen. The image is focussed with the lens in contact with the oil droplet, which greatly enhances the resolution.

Oxidation: the removal of electrons

Parametric test: These are statistical tests that make the assumptions that the values in each population or data set are normally distributed and that they have equal variances. An example is the student's t-test. If there are significant deviations from these assumptions, a non-parametric test should be used.

Peroxisomes: Small single membrane bounded organelles in eukaryotic cells that contain enzymes involved in fatty acid metabolism and breakdown of peroxides. The latter is. catalysed by the enzyme *catalase*, which is enriched in peroxisomes and is used as a biochemical marker enzyme to monitor peroxisome enrichment in cell fractionation procedures.

Peptidoglycan: A covalently cross-linked network of peptides and polysaccharides that make up the structure of bacterial cell walls. It is also referred to as *murein*.

pI: The isoelectric point is the point on a pH scale where the positive charges on a molecule are balanced by the negative charges i.e. no net charge.

pKa: The negative $\log_{10}$ of the equilibrium constant (Ka) i.e. $pKa = -\log_{10} Ka$. The acid and basic forms are in equal concentration

Plasma membrane: a phospholipid and protein fluid bilayer surrounding cells.

P value: The probability that an observed value or mean value could have arisen by chance alone. For example, if two sets of values are compared using an appropriate statistical test and the resultant p value is less than 0.05, there is less than a 5% chance that the differences could have occurred by chance. Therefore the null hypothesis of 'no real difference' is rejected and the two sets of data are considered to be significantly different.

Reduction: the addition of electrons

Reducing agents: A substance (e.g. DTT or 2-ME) that chemically reduces other substances.

Resolving gel: a polyacrylamide gel used to separate complex protein mixture

Ribosomes: Macromolecular structures involved in the synthesis of proteins. They can exist as membrane bound ribosomes (e.g. in the rough endoplasmic reticulum) or 'free' polyribosomes.

Spatula: a small implement with a flat blade used to weigh and mix.

Stacking gel: a polyacrylamide gel which is layered over the resolving gel.

T value: The t value is the value obtained when a t test is performed to compare 2 sets of data. It measures the difference between the means of the two sets or populations of data taking into account the variation between values. Data sets used in such analyses must exhibit a normal distribution.

van der Waals: a weak intermolecular force.

Variance: A measure of dispersion that is equivalent to the mean of the squares of the difference of each value from the population mean. It is the square of the standard deviation.

Zwitterionic: a molecule with both acidic and basic residues e.g. amino acids

ABBREVIATIONS

A_{260nm}:	Absorbance at 260 nm
$\Delta A_{340\ nm}$:	The change in absorbance at 340 nm
A:	Adenine
ATCC:	American Type Culture Collection
ATP:	Adenosine 5'-triphosphate
AMPS:	Ammonium persulphate
Bp:	base pair
C:	Cytosine
C_{18}:	a chain of 18 carbon atoms
C_8:	a chain of 8 carbon atoms
C_5:	a chain of 5 carbon atoms
CHAPS:	(3-[(3-Cholamidopropyl) dimethylammonio]-1-propane sulphonate
CFU:	Colony forming unit
Da:	Dalton
ddA:	dideoxyadenosine triphosphate
ddC:	dideoxycytidine triphosphate
ddG:	dideoxyguanosine triphosphate
ddT:	dideoxythymidine triphosphate
DNA:	Deoxyribonuclec acid
DIC:	Differential interference contrast
DF:	Dye front
2D-PAGE:	Two dimensional polyacrylamide gel electrophoresis
DMSO:	Dimethyl sulphoxide
DTT:	Dithiothreitol
EDTA:	Ethylenediaminetetraacetic acid
$FMNH_2$:	Flavin mononucleotide (reduced)
GLC:	Gas liquid chromatography
G:	Guanine
H & E:	Haematoxylin and eosin
HEPES:	N-2-hydroxyethylpiperazine-N'-2-ethanesulphonic acid
HIC:	Hydrophobic interaction chromatography
HPLC:	High pressure liquid chromatography
IEX:	Ion exchange chromatography
IPG:	Immobilised pH gradient

I.U.:	International units of enzyme activity
Ka:	The equilibrium (association/dissociation) constant.
Kav:	An approximation to K_D in size exclusion chromatography
kDa:	kilo Daltons
K_D:	The equilibrium (distribution or partition) coefficient in chromatography
2-ME:	2-mercaptoethanol
MES:	2(N-morpholine)ethane sulphonic acid
Mr:	Relative molecular mass
MW:	Molecular weight
NaCl:	Sodium chloride
NAD^+:	Nicotinamide adenine dinucleotide (oxidised)
NADH:	Nicotinamide adenine dinucleotide (reduced)
NP:	Normal phase
PAGE:	Polyacrylamide gel electrophoresis
PCR:	Polymerase chain reaction
PFGE:	Pulsed field gel electrophoresis
PFU:	Plaque forming unit
ODS:	Octadecylsilane (C_{18})
PIPES:	1,4-piperazinebis(ethanesulphonic acid)
RCF:	Relative centrifugal force
RER:	Rough endoplasmic reticulum
Rf:	Retention factor (relative mobility or relative to the front)
RNA:	Ribonucleic acid
RP:	Reversed phase
rpm:	Revolutions per minute
S:	Standard deviation (in statistical formulae)
S^2:	Variance
SAX:	Strong anion exchange
SCX:	Strong cation exchange
SD:	Standard deviation
SDS:	Sodium dodecyl sulphate
SDS-PAGE:	Polyacrylamide gel electrophoresis in the presence of SDS
SEC:	Size exclusion chromatography
SEM:	Standard error of the mean (in statistical analysis) or scanning electron microscopy
SER:	Smooth endoplasmic reticulum
SPE:	Sample preparation and extraction
Std:	Standard
TARE:	The weight of the container
TBE:	Tris-Borate-EDTA
TEM:	Transmission electron micrscopy

TEMED:	N,N,N′,N′,-tetramethylethylenediamine
T:	Thymine
TLC:	Thin layer chromatography
Tricine:	N-[tris(Hydroxymethyl)methyl]glycine
Tris:	2-amino-2-hydroxymethylpropane-1,3-diol.
UV:	Ultraviolet light
Ve:	The elution volume of a sample applied to a size exclusion column
Vo:	The void volume of a size exclusion column
Vt:	The total volume of a size exclusion column
(v/v):	Volume in a volume
WAX:	Weak anion exchange
(w/v):	Weight in a volume

1

THE PREPARATION OF SOLUTIONS IN BIOSCIENCE RESEARCH

1.1 Introduction

To help understand the many complex processes and interactions that occur within a cell (*in vivo*) it is often necessary to simulate cellular events in the laboratory (*in vitro*). The many constituents of the cell vary in their concentration, size and physical characteristics and, while it is often difficult to exactly simulate the cellular environment in the laboratory, it is possible to provide an environment in which the cellular constituent under study functions. Then, by careful adjustment of the solution, an understanding of the cellular constituent's role within the cell can be ascertained. This means that the preparation of solutions is a vital step in bioscience research and a topic that all bioscience students should become familiar with.

However, to gain complete control of this topic you *must* memorize key topics, such as the common units used in bioscience research (Table 1.1). The only way to avoid this confusion is to commit to memory the various units involved, or keep a copy of Table 1.1 conveniently placed for ease of use. The units detailed in Table 1.1 are representative of the concentrations in which many metabolites are present within the cell, for example calcium ions (Ca^{2+}) can be present at mM concentrations outside the cell but only nM within the cell's cytoplasm.

Student exercise

- Memorize the units in Table 1.1
- Memorize the sub divisions to the units in Table 1.1.

Having mastered the units, you will be ready to apply the information to the preparation of solutions, although one of the major stumbling blocks you may find is overcoming an aversion to numerical topics. Many students view biosciences as descriptive subjects and try to block out the numerical aspects. If this is the case for you, then this approach will greatly limit your involvement

---

Basic Bioscience Laboratory Techniques: A Pocket Guide, First Edition.
Philip L.R. Bonner and Alan J. Hargreaves.
© 2011 John Wiley & Sons, Ltd. Published 2011 by John Wiley & Sons, Ltd.

Table 1.1 *Common units in bioscience*

	1.0	1.0×10^{-3}	1.0×10^{-6}	1.0×10^{-9}	1.0×10^{-12}
Concentration	Molar (M)	millimolar (mM)	micromolar (μM)	nanomolar (nM)	picomolar (pM)
Amount	mol	millimol (mmol)	micromol (μmol)	nanomol (nmol)	picomol (pmol)
	gram (g)	milligram (mg)	microgram (μg)	nanogram (ng)	picogram (pg)
Volume	Litre (L)	millilitre (ml)	microlitre (μl)	nanolitre (nl)	picolitre (pl)

in the topic. In fact this fear of numeracy is usually misplaced as most of the calculations are straightforward and can become second nature with practice.

On a more practical note, from a student's point of view, numerical assessments within bioscience courses differ from descriptive topics in one notable aspect. In an assessment it is possible to get 100% of the available marks on a numerical question, whereas this is rarely true for a descriptive answer as the marker will always be reluctant to allocate 100% of the marks, signifying a perfect answer.

1.2   Concentration

The concentration of metabolites, chemicals and reagents that are used in bioscience are reported in a few different notations (e.g. molarity and percentage).

1.2.1   Molarity

Science uses molarity to ensure numerical equivalence in terms of the molecules present in a solution. In a mol there are 6.022×10^{23} molecules present (Avogadro's number). This is true for all reagents, for example a mol of sodium chloride (NaCl; Mr 54.55), a mol of glucose ($C_6H_{12}O_6$; Mr 175) or a mol of adenosine triphosphate (ATP; Mr 355) will all have the same number of molecules present. A mol of these reagents will result in different weights because different compounds will have different atoms present in their structure, in effect a different size or mass. But a mol of a reagent will *always* have the same number of molecules present.

Molarity is the most common method of describing the concentration of a reagent in solution. The sentence should be read again. . . 'the *concentration* of a reagent *in solution*'. This describes exactly what occurs; a reagent (possibly a powder) is dissolved in a volume of liquid (usually water). Therefore, two things are required before a concentration can be ascribed; a known weight of a reagent must be physically weighed out, then transferred and dissolved in a known volume of liquid. A mol of a compound is a physical weight (Mr in grammes). You can go to the chemical shelves take the reagent required and weigh out a mol.

When a reagent is dissolved in a volume of liquid the units change from mol to molarity (unit of concentration). The shorthand version of this unit is the capital letter M (Table 1.1). This unit of concentration describes an amount of reagent dissolved in a volume of liquid.

A molar solution (M) is the molecular weight in grams (mol) dissolved in a litre (1000 ml) of water (see Table 1.1).

$$\text{A } 1.0 \text{ M SOLUTION} = \frac{\text{MOLECULAR WEIGHT IN GRAMS (MOL)}}{\text{VOLUME (1 LITRE)}}$$

1.3 Using Balances to Weigh Out Reagents

1.3.1 The Use of an Electronic Balance

In practice the amount of reagent you would weigh out to prepare a solution depends on the availability of the reagent (always check and comply with the Health and Safety data sheet for the reagent before commencing any weighing out) and access to an appropriate accurate balance.

Most bioscience laboratories will have a range of balances accurate to 1, 2, 3 or 4 decimal places. Large amounts of inexpensive reagents are routinely weighed out on 'top pan' balances (see Figure 1.1), whereas small amounts of expensive reagents should be weighed out using an analytical balance. If a hazardous reagent is being weighed out, use a balance contained within a designated safety area with a positive air pressure to ensure that no particles of the reagent can enter the laboratory air space.

Always check to see that the balance is clean. If you need to clean the balance take the appropriate health and safety precautions and dispose of the debris in a prescribed container.

1.3.2 The Use of 'top pan' Balances:

- Place a suitable container (e.g. beaker, plastic weighing boat or aluminium foil) gently onto the top pan balance (see Figure 1.1a) and wait for the

Figure 1.1 *Examples of typical balances (a) top pan (b) analytical.*

electronic readout to stabilize. Press the zero button (or TARE) on the balance to bring the reading to zero.
- Using an appropriate spatula (see Figure 1.2a), gently add the reagent to the container to achieve the correct weight.

Then remove the container and ensure the balance is clean for the next user.

1.3.3 The Use of Analytical Balances

Analytical balances are sensitive to surrounding vibrations and are usually placed on a vibration free bench (see Figure 1.1b).

- Check that the balance is clean and if it requires cleaning take the appropriate health and safety precautions.
- Open the side door to the balance and place a suitable container on the balance pan (e.g. small beaker, a plastic tube, plastic weighing boat or aluminium foil).
- Close the door and allow the electronic readout to stabilize. Press the zero button (or TARE) on the balance to bring the reading to zero. Alternatively, note the weight of the beaker and add the weight of the beaker to the amount of reagent required.
- Add the reagent to the container on the pan of the analytical balance using a micro spatula (see Figure 1.2b), close the side door and allow the read out to stabilize. To weigh out accurately is a skill, but with practice it is possible to weigh out as little as 1.0 mg.
- When the desired amount has been weighed out the container should be carefully removed and the reagent dissolved.

Figure 1.2 A typical (a) spatula and (b) micro spatula.

- Remember to check that the floor and the pan of the balance are clean. Usually there are brushes and wipes available to keep these important devices clean (remember to dispose of any debris in compliance with health and safety regulations).

1.4 Practical Considerations When Making a 1.0 M Solution

No matter how careful you are in preparing a solution for use in bioscience research, there is always the possibility that bacteria or fungi may be present initially in small numbers in the final solution. This may not be a problem if the solution is to be used immediately, but it can be a problem if the solution is left in storage for a period of time.

- Storing solutions at a reduced temperature will arrest or slow down bacterial growth. A solution that is in regular use should be stored in a refrigerator (4–8 °C) and then warmed to the required operating temperature before use. Stocks of solutions can be stored frozen (−25 °C) in convenient aliquots, these can then be thawed and warmed to the required operating temperature before use.
- A concentrated stock solution (e.g. ×10) may help prevent (or arrest) the growth of bacteria or fungi. Dilutions can be prepared from the stock which will take up less storage space.
- If the reagent is heat stable the prepared solution can be placed in a suitable container and heated to 120 °C under pressure in an autoclave. Allow the solution to cool before storage.
- Alternatively, before storage, the solution can be passed through a filter with a pore size (0.2 μm or 0.45 μm) which is small enough to trap bacteria.
- Another approach involves using aseptic techniques to prepare the solution in a filtered air stream (see Chapter 8 for further details).
- Different combinations of the above may be required.

Working out the amount required to make a 1.0 M solution of any reagent is relatively straight forward. You simply find the relative molecular mass (Mr) from the reagent bottle or from a reference book, accurately weigh the required amount, transfer the solid to a suitable beaker (or vessel) then add the highest quality water available[1] initially to a volume less than a litre to dissolve the reagent (see worked example 1.1).

Dissolving a reagent completely is required for the correct final concentration to be achieved. It is usual to add a magnetic stirring flea into the solution in a beaker and then place this on a metallic stirring device. Set the device to a slow stir and wait for the reagent to dissolve (see Figure 1.7). Difficult to dissolve

Figure 1.3 *Volumetric (a) flasks and (b) pipettes.*

reagents may require the addition of heat and hot plate stirrers can be used in conjunction with a glass beaker (remember to use the appropriate safety precautions). Alternatively, a microwave oven can be used to slightly elevate the temperature of the liquid. In either case boiling the water should be avoided. Ultrasonic baths can be used to quickly disperse solids into a solution but remember that prolonged use of these baths will eventually elevate the temperature of the solution.

If the pH of the reagent requires (see Section 1.7.1) adjusting, then the solution should be first cooled to the temperature at which it is to be used (the reason for this is that some reagents have a different pH at different temperatures). The pH can be adjusted (see Figure 1.7) with the addition of the appropriate acid or base before the solution is added to a volumetric flask (see Figure 1.3a).

Worked Example 1.1: Prepare a 1.0 M solution of NaCl

- 1.0 M solution of NaCl will have 54.5 g of NaCl in 1000 ml.
- Weigh out 54.5 g of NaCl in a 1000 ml beaker on a top pan balance.
- Add 750 ml of distilled water to dissolve the NaCl. This can be helped by the addition of a stirring flea and then placing the beaker on a magnetic stirring device.

- When the solid has dissolved transfer the liquid to a 1000 ml volumetric flask. Volumetric flasks have a mark etched on the neck of the vessel, when the meniscus of the liquid touches this mark the correct volume is achieved.
- Add distilled water to the flask until the meniscus of the liquid is at the mark on the neck of the volumetric flask.
- Add a stopper and invert the flask 5–10 times to mix the liquid.
- Transfer the liquid in the volumetric flask to storage bottle/flask. Label this with (a) the name of the reagent, (b) the molarity, (c) the pH, (d) any specific health and safety precautions (e.g. corrosive or poisonous), (e) the date of preparation and (f) your own name and course.
- Store the liquid in the fridge or at room temperature.

1.4.1 Preparing Solutions with a Concentration Less than 1.0 M and a Volume Less than 1.0 Litre

The concentration of metabolites within the cell is rarely in the molar range, so less concentrated solutions are often required for convenience and to reduce the costs. It is these calculations that open up a variety of errors for many students, but if you break the calculation down to its constituent parts and practice, the calculation will become second nature.

Worked Example 1.2: Prepare 25 ml solution of adenosine triphosphate (ATP) at a final concentration of 50 mM

- *Write down what weight of ATP would be required for a 1.0 M solution*
507.2 g of ATP would need to be weighed out and dissolved in 1.0 L (1000 ml) to make a 1.0 M solution. Clearly with expensive reagents this is a prohibitive amount.
- *Adjust for the volume required.*
We don't want 1000 ml (1.0 litre), we only require 25 ml (1.0 ml is 10^{-3} of 1.0 litre). So divide the Mr of ATP by 1000 to get to 1.0 ml and then multiply by 25 to get to the required volume. This is now the weight in grams of ATP required to make 25 ml of a 1.0 M solution.

$$\frac{507.2 \times 25}{1000} = 12.68 \, \text{g} \dots (A)$$

We are preparing a solution which is in a smaller volume than a litre so always check that the weight calculated is less than the Mr of the required reagent.
- *Adjust for the concentration.*
We don't want 25 ml of a 1.0 M solution, we require a final concentration of 50 mM (1.0 mM is 10^{-3} of 1.0 M). So divide the weight of ATP obtained

above (A) by 1000 to get to 1.0 mM and then multiply by 50 to get to the required molarity.

$$\frac{12.68 \times 50}{1000} = 0.63 \text{ g} \dots (B)$$

This is the weight (0.63 g) of ATP required to make 25 ml of a 50 mM solution.

This may seem a basic piece of advice for graduate students, but everyone is capable of making a mistake in these calculations and if you stick to this method you should avoid many errors.

There are sometimes requirements for less concentrated solutions. The series of calculations outlined above can be used for these less concentrated solutions as well.

Worked Example 1.3: : Prepare 50 ml solution of ATP at a final concentration of 750 μM

• *Write down what weight of ATP would be required for a 1.0 M solution*
507.2 g of ATP would need to be weighed out and dissolved in 1.0 L (1000 ml) to make a 1.0 M solution. Clearly with expensive reagents this is a prohibitive amount.
• *Adjust for the volume required.*
We don't want 1000 ml (1.0 litre) we require 50 ml (1.0 ml is 10^{-3} of 1.0 litre). So divide the Mr of ATP by 1000 to get to 1.0 ml and then multiply by 50 to get to the required volume. This is now the weight in grams of ATP required to make 50 ml of a 1.0 M solution.

$$\frac{507.2 \times 50}{1000} = 25.36 \text{ g} \dots (A)$$

We are preparing a solution which is in a smaller volume than a litre so always check that the weight calculated is less than the Mr of the required reagent.
• *Adjust for the concentration.*
We don't want 50 ml of a 1.0 M solution, we require a final concentration of 750 μM (1.0 μM is 10^{-6} of 1.0 M). So divide the weight of ATP obtained above (A) by 1 000 000 to get to 1.0 μM and then multiply by 750 to get to the required molarity.

$$\frac{25.36 \times 750}{1000 \times 1000} = 0.019 \text{ g} \dots (B)$$

This is the weight (0.019 g) of ATP required to make 50 ml of a 750 μM solution.

Using this method it is possible to correctly calculate the amount in grams of any reagent required to make up any concentration or volume.

1.4.2 Preparing Percentage Solutions

In some circumstances percentage solutions are used in bioscience research and these solutions have different units.

A percentage solution means that an amount of reagent in grams is dissolved for every 100 ml of liquid. When the percentage unit is (w/v) it is a weight of a reagent in a 100 ml volume (w/v) (e.g. 0.9% w/v NaCl).

This means that if 100 ml of this solution is required, it is prepared by weighing out 0.9 g of NaCl into a suitable vessel. A volume of water less than 100 ml is added and the NaCl is dissolved. When the NaCl has dissolved the solution can then be made up to 100 ml to give the final percentage solution of 0.9% (w/v) NaCl.

If a different final volume is required then both the amount of reagent and volume should be adjusted accordingly.

Worked Example 1.4: : The preparation of % solutions

• *Prepare 25 ml of 0.9% (w/v) NaCl*
Adjust for the volume by dividing by 100 to get the amount required for 1.0 ml and then multiply by 25 to get the amount required for 25 ml.

$$= \frac{0.9 \times 25}{100} \, g$$

$$= 0.225 \, g$$

Dissolve this amount of reagent in a volume less than 25.0 ml; when this has dissolved, make the volume up to 25.0 ml.

• *Prepare 375 ml of 0.9% (w/v) NaCl*
Adjust for the volume by dividing by 100 to get the amount required for 1.0 ml and then multiply by 375 to get the amount required for 375 ml.

$$= \frac{0.9 \times 375}{100} \, g$$

$$= 3.375 \, g$$

Dissolve this amount of reagent in a volume less than 375.0 ml; when the reagent has dissolved make the volume up to 375.0 ml.

Liquids can also be weighed out to give a weight in a volume solution for example 2% (w/v) glycerol. If 100 ml of 2% (w/v) glycerol was required, 2.0 g of the liquid glycerol would be weighed into a beaker and water added to a volume less than 100 ml. When the glycerol had dissolved the volume could then be made up to 100 ml.

However, liquids are frequently prepared as a volume within a volume concentration (v/v) (e.g. 5% v/v methanol). If 100 ml of 10% (v/v) methanol solution is required, 10 ml of methanol is measured out and made up to 100 ml with distilled water. Smaller volumes of the same percentage can be prepared by adjusting the volumes of both liquids proportionally.

1.4.2.1 *Weight in a Volume (w/v) Solutions*

Some common materials encountered in bioscience, for example proteins and nucleic acids are commonly dissolved as a weight in a declared volume for example mg ml^{-1} (this can also be written as mg/ml). The reason for this is that solutions of proteins can be a heterogeneous population of different molecules, which makes constructing molar concentrations meaningless. The concentrations of proteins from cellular extracts are commonly estimated using reagents which interact with proteins to produce a coloured product (see Chapter 3.8).

1.5 Dilutions and the Use of Pipettes

It was proposed (Section 1.4) that a convenient method to prepare solutions of different concentrations is to initially prepare a concentrated stock solution and then dilute down from this concentrated solution to the concentration and volume required. The ability to prepare accurate dilutions is an important skill in bioscience research (also see Chapters 2 and 8).

Preparing large volume dilutions can be achieved using a measuring cylinder (see Figure 1.4a) but accurate dilutions requires the use of a pipette. Volumetric glass (see Figure 1.3b) or cylindrical graduated glass/plastic pipettes (see Figure 1.4b) are still in use but these have been superseded by the use of mechanical (or electronic) micropipettes (see Figure 1.5), primarily on health and safety issues, accuracy and ease of use. All pipettes work by creating a vacuum above the liquid that requires dispensing in a liquid holding barrel. The liquid is dispensed by releasing the vacuum with care.

KEY POINTS TO REMEMBER 1.1.

- A dilution implies that the concentration of the final working solution is *less* than the concentration of the stock solution. If your calculation produces a more concentrated solution than the stock you must start the calculation again. It is also important to ensure that the units of volume and concentration (see Table 1.1) are the same on both sides of the equation.

- A 1 in 10 dilution is 1 part in a final volume of 10. (*Not* one part added to ten parts).

Figure 1.4 *An example of (a) a measuring cylinder, (b) cylindrical pipette and (c) pipette filler.*

1.5.1 The Use of Graduated Cylindrical Glass or Plastic Pipettes

If a graduated cylindrical glass or plastic pipette is to be used to prepare a dilution from a stock solution:

- Never use your mouth to draw liquid into the pipette!!
- Ensure that the pipette is clean and has no flaws at either end.

- *Hold the pipette at the top* and insert it into the nozzle of a pipette filler (see Figure 1.4c or Figure 8.1) using a firm hold. Do not try and force the pipette into the filler; if there is an obviously poor grip between the filler and the pipette then exchange the pipette filler for a new one.
- Draw liquid into the cylindrical barrel to the level required as recommended by the manufacturer. Hold the pipette above the stock liquid and check to see if the pipette drips liquid. If the pipette drips liquid replace the pipette (or the pipette filler/or both).
- Dispense the stock solution into a suitable vessel.
- Place the used pipette in a suitable tray for cleaning.

Figure 1.5 *Example of two common laboratory pipettes.*

1.5.2   The Use of Hand Held Micropipettes

Micropipettes are routinely used in bioscience laboratories and, although there are many different manufacturers of hand held micropipettes, the design is inherently similar (see Figure 1.5).

- Micropipettes are accurate only between the range of volumes designated by the manufacturer (e.g. 100–1000 µl). This range is usually on the side and top of the pipette, so check to see if the pipette covers the range required. If it doesn't, then change the pipette.
- Adjust the volume on the pipette to the volume required (remember that 1.0 ml is 1000 µl – see Table 1.1). Do not adjust the pipette beyond its designated range.
- Place a suitable tip on the end of the pipette (pipettes with different volume ranges require different tips. Please check if you are unsure).
- Before drawing up the liquid, hold the pipette in your hand and depress the plunger using the thumb. Notice that there are two stops to which the plunger can be depressed; the first stop on the pipette will allow the required volume of

13

liquid to be taken up into the pipette tip and the second stop allows any trace liquid to be expelled from the pipette tip.

- With the pipette tip on the end of the micropipette depress the plunger to the first stop; place the tip into the liquid, slowly allowing the plunger to return to the starting position. Check to see that no air bubbles have been drawn into the tip.
- Move the pipette to the dispensing container and depress the plunger initially to the first stop, this is followed by moving the plunger to the second stop.
- The pipette tip can then be removed (most micropipettes have a mechanical device to remove the tip) and disposed of according to health and safety regulations.
- Your ability to pipette accurately can be practiced using an empty beaker on a balance and some distilled water. Set the pipette to a volume and dispense water into the beaker on the balance. Note the weight at each addition and after ten additions work out the average weight and the standard deviation. The average weight should correspond to the setting on the pipette which will indicate the accuracy of the pipette and the standard deviation will show you how adept you are at dispensing the same volume.

KEY POINTS TO REMEMBER 1.2

- 0.1 ml of water weighs 0.1 g (100 mg)
- 1.0 ml of water weighs 1.0 g
- 1000 ml weighs 1.0 kg.

1.5.3   Diluting Down from a Stock Solution

A convenient formula can be used to calculate dilutions from a stock solution

$$C_1V_1 = C_2V_2$$

Initial concentration (C_1) x initial volume (V_1) = final concentration (C_2) x final volume (V_2)

Worked Example 1.5:

- *You are supplied with 1.0 M stock solution of pyruvate.*
What volume of the 1.0 M pyruvate stock is required if you need to prepare 5 ml of a 25 mM solution?

$$C_1V_1 = C_2V_2$$

$1000 \times V_1 = 5 \times 25 \, (\text{note } C_1 \text{ is 1000 mM} - \text{see Table 1.1})$

$$V_1 = \frac{5 \times 25}{1000} \, \text{ml}$$

$$= 0.125 \, \text{ml (note this is also 125 } \mu\text{l—see Table 1.1)}$$

Take 0.125 ml of the 1.0 M stock solution and make this up to 5.0 ml to arrive at a 25 mM solution.

• *You are supplied with a 25 mM stock solution of L-lysine.*

What volume of the 25 mM L-lysine stock is required if you need to prepare 0.75 ml (750 μl) of a 500 μM solution?

$$C_1 V_1 = C_2 V_2$$

Convert all the units to the 1×10^{-3} level (milli) (see Table 1.1)

$$25 \times V_1 = 0.75 \times 0.5 \text{ (note } C_1 \text{ is 0.75 mM} - \text{see Table 1.1)}$$

$$V_1 = \frac{0.75 \times 0.5 \, \text{ml}}{25}$$

$$= 0.015 \, \text{ml (note this is also 15} \mu\text{l} - \text{see Table 1.1)}$$

Answer: Take 0.015 ml (15 μl) of the 25 mM stock solution and make this up to 0.75 ml to arrive at a 500 μM solution.

Alternatively:

• *Convert all the units to 1×10^{-6} level (micro) (see* Table 1.1)

$$25,000 \times V_1 = 750 \times 500 \text{(note } C_1 \text{ is 25 mM or 25000 } \mu\text{M} - \text{see Table 1.1)}$$

$$V_1 = \frac{500 \times 750}{25,000} \, \mu\text{L}$$

$$= 15 \, \mu\text{l (note this is also 0.015 ml see Table 1.1)}$$

Take 15 μl of the 25 mM stock solution and make this up to 750 μl (0.75 ml) to arrive at a 750 μM solution.

1.6 Water, Acids and Bases

Most living matter is composed of approximately 70% water and many of its properties, including the formation of H-bonds and the ability to ionize, are fundamental to the correct structure and properties of many biological compounds. Understanding the properties of water and how it interacts with other molecules is fundamental to understanding many topics within biosciences.

The structure of water means that there are electron rich areas (enriched negative charge) within the molecule and electron depleted areas (enriched positive charge) called a charge dipole. This dipole of charge contributes many of the properties of water because the dipole on one molecule of water can interact with the dipole on other water molecules (see Figure 1.6a), thus establishing a network of interactions described as hydrogen bonds (H-bonds). In water most of the molecules are engaged in hydrogen bonding with their nearest neighbours and each bond lasts between 1–20 pico seconds (1 picosecond $= 10^{-12}$ seconds) before the bond is reformed or a new bond is formed with a neighbouring water molecule. Hydrogen bonds are relatively weak bonds (the *bond dissociation energy* is low 23 kJ mol^{-1}) if this is compared to some covalent bonds (470 kJ mol^{-1} to break the O=H covalent bond or 348 kJ mol^{-1} to break a C=C covalent bond). This is also reflected in the greater bond distance of the hydrogen bond compared to a covalent bond (see Figure 1.6a). This means that, as the temperature rises (hence an increase in available energy), the amount of H-bonding decreases. In ice, H_2O molecules form four hydrogen bonds with other H_2O molecules to form a low temperature crystal lattice. The elevated temperature of water results in an increase in the random movement of the H_2O molecules and each molecule hydrogen bonds with only 3.4 water molecules.

KEY POINTS TO REMEMBER 1.3

- Hydrogen bonds are important weak bonds that hold the structures of proteins and nucleic acids together. In most cases elevated temperatures (e.g. $> 37\,^{\circ}$C) will alter the structure of biological macromolecules.

- Hydrogen bonding can be formed between molecules other than water in biological systems. The dipole on electron rich atoms such as oxygen and nitrogen (relatively negative and thus a hydrogen bond acceptor) and the dipole on the hydrogen attached to oxygen or nitrogen (relatively positive and thus a hydrogen bond acceptor) may interact if they are within the correct bond forming distance.

1.6.1 Water as a Solvent

Water is a polar solvent because the dipole on water molecules enables water to interact with itself and other charged (*polar*) molecules, allowing them to dissolve (*hydrophilic*) (see Figure 1.6b) compounds. In contrast, non polar (*hydrophobic*) molecules cannot interact with water, and in the presence of water non polar molecules tend to group together. The water molecules surrounding the non polar molecules form caged (*clathrate*) structures (see Figure 1.6c).

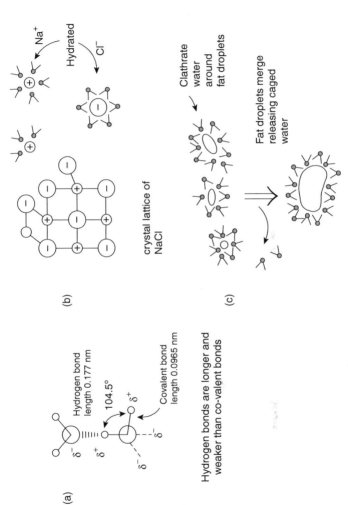

Figure 1.6 The structure of water (a) contributes to its ability to solubilise ionic compounds (b) but not non ionic compounds (c).

17

1.6.2   The ionization of water

For every billion water molecules (10^9) there are approximately two molecules that ionize at room temperature. This of course is a very low degree of ionization but this still represents another important property of water.

Water ionizes into a positively charged hydrogen ion [H^+]* and a negatively charged hydroxyl ion [OH^-]**

$$H_2O \Leftrightarrow [H^+] + [OH^-]$$

All reactions can reach equilibrium where the rate of the dissociation is equalled by the rate of association. The equilibrium constant for water Keq at 25 °C is given by the equation.

$$Keq \Leftrightarrow \frac{[H^+][OH^-]}{[H_2O]} \qquad (1.1)$$

In pure water the molar concentration of water is 55.5 M (Divide the weight of 1.0 litre of water [i.e. 1000 g] by the relative molecular mass of water; Mr = 18) and the Keq for water has been determined experimentally to be 1.8×10^{-16}.

Substitute these values into Equation (1.1).

$$1.8 \times 10^{-16} \Leftrightarrow \frac{[H^+][OH^-]}{[55.5]}$$

The ion product of water (Kw) at 25 °C is

$$Kw = [1.8 \times 10^{-16}] [55.5] = [H^+][OH^-]$$

$$Kw = 1 \times 10^{-14} M$$

The ion product of water (Kw) always equals 1×10^{-14} M. When the concentration of [H^+] is equal to the concentration of [OH^-] the solution is at the neutral pH value (7.0).[2]

This value can be calculated when the concentration of [H^+] is equal to the concentration of [OH^-].

$$Kw = [H^+][OH^-] = [H^+]^2$$

Solving this for [H^+]

$$[H^+] = \sqrt{Kw} = \sqrt{1 \times 10^{-14}}$$
$$= 1 \times 10^{-7}$$

The ion product of water is constant which means that when the concentration of [H^+] is high, there is a corresponding decrease in the concentration of [OH^-].

Table 1.2 *The pH scale*

[H^+] (M)	pH	[OH^-] (M)
10^0 (1.0)	0.0	10^{-14} (0.00000000000001)
10^{-1} (0.1)	1.0	10^{-13} (0.0000000000001)
10^{-2} (0.01)	2.0	10^{-12} (0.000000000001)
10^{-3} (0.001)	3.0	10^{-11} (0.00000000001)
10^{-4} (0.0001)	4.0	10^{-10} (0.0000000001)
10^{-5} (0.00001)	5.0	10^{-9} (0.000000001)
10^{-6} (0.000001)	6.0	10^{-8} (0.00000001)
10^{-7} (0.0000001)	7.0	10^{-7} (0.0000001)
10^{-8} (0.00000001)	8.0	10^{-6} (0.000001)
10^{-9} (0.000000001)	9.0	10^{-5} (0.00001)
10^{-10} (0.0000000001)	10.0	10^{-4} (0.0001)
10^{-11} (0.00000000001)	11.0	10^{-3} (0.001)
10^{-12} (0.000000000001)	12.0	10^{-2} (0.01)
10^{-13} (0.0000000000001)	13.0	10^{-1} (0.1)
10^{-14} (0.00000000000001)	14.0	10^0 (1.0)

The ion product of water (Kw) is the basis of the pH scale, which denotes the molar concentration of [H^+] in solution.

$$pH = -\log_{10}[H^+]$$

By taking the negative logarithm the exponential values (1×10^{-X}) values seen above are converted into linear arithmetic numbers (e.g. at neutral pH the [H^+] = 1×10^{-7} M which is converted to pH 7.0 (see Table 1.2).

- The pH scale is the negative logarithm of the hydrogen ion concentration [H^+]. This means that the values for the molar concentration of [H^+] decrease by a factor of 10 with each pH number not by a factor of 1.

1.6.3 Acids and Bases

The pH scale outlined in the previous section is a measure of the acidity of a solution describing the molar concentration of [H^+] in solution. The compounds which can release H^+ ions into solution are known as acids. In 1923 two scientists Johannes Nicolaus Brønsted (Denmark) and Thomas Martin Lowry

(England) independently proposed essentially the same theory of acids and bases. This is now known as the Brønsted-Lowry theory of acids and bases, which proposes that an acid is a proton (H^+) donor and a base is a proton (H^+) acceptor. In addition, every acid has a conjugate base and every base a conjugate acid.

$$\text{e.g.} \quad \underset{\text{acid}}{HCl} + \underset{\text{base}}{H_2O} \quad \leftrightarrow \quad \underset{\text{acid}}{H_3O^+} + \underset{\text{base}}{Cl^-}$$

At the same time, another scientist Gilbert Newton Lewis (USA) proposed that acids can be considered as electron pair donors (Lewis acid) and bases as electron pair acceptors (Lewis base).

1.6.3.1 *Strong and Weak Acids*

Acids are often referred to as being 'strong' or 'weak'. This does not refer to their ability to dissolve other compounds, but informs us of their ability to ionize and dissociate in solution.

A strong acid (e.g. hydrochloric acid (HCl), nitric acid (HNO_3) or sulphuric acid (H_2SO_4) will ionize and dissociate at *all pH values* into their conjugate parts of acid and base.

i. $\underset{\text{acid}}{HCl} + \underset{\text{base}}{H_2O} \quad \leftrightarrow \quad \underset{\text{acid}}{H_3O^+} + \underset{\text{base}}{Cl^-}$

ii. $\underset{\text{acid}}{HNO_3} + \underset{\text{base}}{H_2O} \quad \leftrightarrow \quad \underset{\text{acid}}{H_3O^+} + \underset{\text{base}}{NO_3^-}$

iii. $\underset{\text{acid}}{H_2SO_4} + \underset{\text{base}}{2H_2O} \quad \leftrightarrow \quad \underset{\text{acid}}{2H_3O^+} + \underset{\text{base}}{SO_4^{2-}}$

In contrast, the degree of ionization and dissociation of 'weak acids' is dependent upon the pH of the solution in which they are dissolved. Acetic acid (CH_3COOH) is a familiar weak acid which ionizes to varying degrees across the pH scale (see Figure 1.7).

$$CH_3COOH + H_2O \quad \leftrightarrow \quad H_3O^+ + CH_3COO^-$$

Below pH 3.0, when the $[H^+]$ concentration is high (see Table 1.2), any loss of the $[H^+]$ on the carboxyl group of acetic acid ($=COOH$) is quickly replaced. However, above pH 3.0 (i.e. when the $[H^+]$ concentration in solution decreases rapidly; see Table 1.2) the proton on the carboxyl group of acetic acid can dissociate and go into solution. This leaves a negatively charged acetic acid and a positively charged proton in solution.

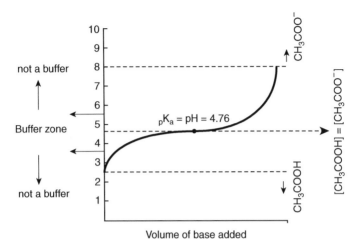

Figure 1.7 A titration curve for acetate buffer.

1.7 Buffers

Many biological compounds have weak acidic groups present within their structure. As a result, the surface charge on the molecule will vary according to the surrounding pH. Amino acids, proteins, nucleotides and nucleic acids have weak acidic groups as part of their structures and the degree of ionization of these molecules is pH dependent. As the integrity and functionality of these molecules depends on the pH, maintaining this optimum pH value is *extremely important* in biology.

1.7.1 Preparing a Buffer and Using a pH Meter

There are many different buffers used in bioscience laboratories to preserve the pH of a solution (some listed in Table 1.3). A buffer comprises of a weak acid and one of its salts (conjugate base) or a weak base and one of its salts (conjugate acid). Buffers work best when there is a balance between charged and uncharged species (pKa) (see Figure 1.7 and section 1.8). At the pKa buffers are able to absorb changes in the concentration of $[H^+]$ protons or hydroxyl groups $[OH^-]$ without significantly altering the overall pH. There is a limit to this buffering capacity which is approximately 1.0 pH unit either side of the buffer's pKa. If the

Table 1.3 *A list of common buffers and their effective pH range*

Buffer	pKa	Effective buffer range
Maleate	1.97	1.2–2.6
Acetate/CH_3COOH	4.76	3.6–5.6
MES	6.10	5.5–6.7
PIPES	6.76	6.1–7.5
NaH_2PO_4/Na_2HPO_4	7.20	5.8–8.0
HEPES	7.48	6.8–8.2
Tricine	8.05	7.4–8.8
Tris/HCl	8.06	7.5–9.0
Na_2CO_3/$NaHCO_3$	10.33	9.0–10.7

desired pH value is at the extremes of the buffer's pKa value, the molarity of the buffer should be increased or a different buffer used.

- When the desired molarity of the buffer has been decided, the buffer is weighed out and dissolved (see Section 1.4) in a volume of water (highest purity available) less than the final volume required. Any additions to the buffer which may contribute to the pH value (e.g. EDTA) should be dissolved at this stage before the pH is finalized. When the buffer has dissolved, the pH can be adjusted using a pH meter (see Figure 1.8).
- The probe of the pH meter should be cleaned by rinsing with distilled water and the machine's response calibrated with standard buffer solutions according to the manufacturer's instructions.
- The pKa value for some buffers can vary with changes in the temperature (e.g. pKa of Tris at 4 °C is 8.8 but this will change to 8.3 at 20 °C). Most pH meters will also have a temperature adjustment switch. Consider equilibrating the buffer to the temperature it is to be used at in an experiment before adjusting the pH.
- The buffer solution can be placed on a magnetic stirring plate and a suitable magnetic stirring 'flea' should be rinsed before placing in the buffer solution. The stirring rate should not be too fast as the flea may jump and damage the pH probe (particularly important when using glass probes).
- The probe should be rinsed with distilled water prior to being placed in the buffer solution.

Figure 1.8 *A typical set up to adjust the pH of a solution.*

- Allow a few minutes for the pH of the solution to stabilize on the pH meter's readout before starting to adjust the pH.
- Remember that buffers are made up of predetermined combinations; for example sodium acetate buffer comprises a basic component (sodium acetate) and an acidic component (acetic acid). If the desired pH value is overshot, avoid adjusting the pH back to the correct pH with a strong acid or base, for example hydrochloric acid (HCl) or sodium hydroxide (NaOH). This will use up the buffering capacity creating a less effective buffer. Instead, adjust to the correct pH using a concentrated solution of either one of the buffer's constituents; this will alter the molarity of the buffer but retain the buffering capacity.
- When the correct pH has been attained the pH probe can be removed and rinsed in distilled water before being returned to the storage solution.
- The buffer can then be made up to volume and stored as required. If the buffer is to be used for chromatography (see Chapter 7) it is advisable to filter the buffer through a $0.2\,\mu m$ filter to remove particulate matter and bacteria.
- Overnight storage in a 0.1 M HCl solution will clean and restore most dirty pH probes.

1.8 The Equilibrium/Dissociation Constant (Ka) for an Acid or Base and the Henderson –Hasselbalch Equation

A measure of the strength of an acid is the acid-dissociation equilibrium/dissociation constant (Ka) for that acid.

$$\underset{\text{acid}}{HA} + \underset{\text{base}}{H_2O} \quad \leftrightarrow \quad \underset{\text{acid}}{H_3O^+} + \underset{\text{base}}{A^-}$$

The dissociation constant $Ka = \frac{[H_3O^+][A^-]}{[HA]}$

- A strong acid readily dissociates at all pH values and has a high Ka value (HCl: $Ka = 1 \times 10^3$).
- A weak acid has a small Ka value (see Figure 1.7) and its dissociation is pH dependent (CH_3COOH $Ka = 1.8 \times 10^{-5}$).
- To convert Ka values into useable numbers, the pKa of an acid or base is used which is the negative logarithm of the Ka. In this case a strong acid will have a low pKa.

$$pKa = \log\frac{1}{Ka}$$

The Henderson-Hasselbalch equation.

The quantitative aspects of buffers can be ascertained by the Henderson-Hasselbalch equation, which mathematically links pH with pKa.

$$pH = pKa + \log 10\frac{[A^-]}{[HA]} \quad \text{or} \quad pH = pKa + \log 10\frac{[\text{proton acceptor}]}{[\text{proton donor}]}$$

- Where $[A^-]$ is the concentration of base and $[HA]$ is the concentration of acid. This is usually a molar concentration but because $[A^-]/[HA]$ is a ratio, other concentration units will also be acceptable. It is this ratio which determines the pH of a solution.
- At the mid point of the titration curve (see Figure 1.7) when the concentration of $[HA]$ is equal to the concentration of $[A^-]$ the pH = pKa and this is when buffers are most effective.

$$pH = pKa + \log 1 = pKa + 0 = pKa$$
$$\text{or } pH = pKa$$

When $[HA] > [A^-]$ the pH of the buffer is less than the pKa and when the $[A^-] > [HA]$ the pH is greater than pKa.

- This $[A^-]/[HA]$ ratio can only be varied within certain limits usually to 1.0 pH unit either side of the pKa value. This means that there is little or no buffering at the extremes of the buffering range and it may be necessary to increase the concentration of the buffer to maintain good buffering capacity or switch to a different buffer with a different pKa.

The Henderson – Hasselbalch equation allows

- Calculation of pKa when the pH and molar concentrations of proton donor and acceptor are known.
- Calculation of the pH if the pKa and molar concentrations of proton donor and acceptor are known.
- Calculation of the molar concentrations of proton donor and acceptor if the pKa and pH are known.

Worked Example 1.6: The use of the Henderson–Hasselbalch equation

Calculate the pH of a solution containing 0.15 M acetic acid and 0.25 M sodium acetate. (For the constituents of a sodium acetate buffer see Table 1.3).
   The pKa of acetic acid is 4.76

$$\underset{\text{acid}}{HA} + \underset{\text{base}}{H_2O} \quad \leftrightarrow \quad \underset{\text{acid}}{H_3O^+} + \underset{\text{base}}{A^-}$$

Acetic acid   acetate
$$\underset{\text{acid}}{CH_3COOH} + \underset{\text{base}}{H_2O} \quad \leftrightarrow \quad \underset{\text{acid}}{H_3O^+} + \underset{\text{base}}{CH_3COO^-}$$

$$pH = pKa + \log_{10} \frac{[A^-]}{[HA^+]} \quad \text{or} \quad pH = pKa + \log_{10} \frac{[\text{proton acceptor}]}{[\text{proton donor}]}$$

$$pH = pKa + \log_{10} \frac{[\text{acetate}]}{[\text{acetic acid}]}$$

$$pH = 4.76 + \log_{10} \frac{[0.25]}{[0.15]}$$

$$pH = 4.76 + 0.22$$

$$pH = 4.96$$

1.9 Summary

The foundation of successful experimental work is the correct preparation of solutions. If undertaking a course in the bioscience area the following skills must be developed to a satisfactory standard, particularly if you wish to pursue a career in laboratory based bioscience.

- Preparation of molar and percentage solutions.
- Preparation of dilutions prepared from a stock solution.
- The accurate use of pipettes.
- A good understanding of pH and the preparation and use of buffers.

Notes

[1] Water from a tap may contain impurities (such as metal ions) which may interfere with an experiment. Water obtained by distillation represents the minimum purity of water that ideally should be used to prepare solutions. Water purification machines which pump tap water through filters can be used to produce distilled and deionized water of varying degrees of purity. The highest purity of water will have a conductivity of 18.2 Ω ms^{-1} (0.055 μS).

[2] The pH of the cell is approximately 7.0. This is also called the physiological pH value. Water is rarely at pH 7.0 because carbon dioxide (CO_2) dissolves in it to produce carbonic acid, resulting in a slightly acid pH for water of approximately pH 5.5.

$$CO_2 + H_2O \Leftrightarrow H_2CO_3 + H_2O \Leftrightarrow \underset{acid}{H_3O^+} + \underset{base}{HCO_3^-}$$

[*] [H$^+$] is effectively a proton and this is never available in solution. The proton joins up with another water molecule to generate a positively charged hydronium ion [H$_3$O$^+$].

[**] [OH$^-$] the square brackets indicate that a defined concentration is involved and this is usually molar concentration.

2

MICROSCOPY

2.1 Introduction

The major differences between prokaryotes and eukaryotes are the presence of a nuclear envelope, a cytoskeleton network and intracellular organelles in the latter. Most bacteria (prokaryotes) have a peptidoglycan cell wall for structural stability and a membrane that encloses the cytoplasm, whereas the plant and yeast (eukaryotes) cell walls are made up of other components such as cellulose and chitin, respectively. Animal cells have no cell wall whatsoever (Figure 2.1).

Organelles such as nuclei, mitochondria, lysosomes, peroxisomes, rough and smooth endoplasmic reticulum and Golgi apparatus are present in all eukaryotes, but not in prokaryotes, whereas plants have additional intracellular structures not present in yeasts or animal cells. These are summarized in Figure 2.1 and in the bullet point lists below:

The principal organelles/compartments in animal cells and their main functions are:

- Nucleus: DNA replication and transcription.
- Endoplasmic reticulum: Synthesis of membrane proteins and membrane lipids, Ca^{2+} storage and detoxification of xenobiotics.
- Lysosomes: Degradation of macromolecules by acid hydrolases.
- Peroxisomes: detoxification of hydrogen peroxide.
- Mitochondria: ATP production and Ca^{2+} storage. Semi-autonomous organelles that synthesize some of their own proteins. Mitochondrial DNA and ribosomes are located in the mitochondrial matrix. Double membrane structure with inner membrane folded to form cristae.
- Free polyribosomes: Synthesis of cytoplasmic, nuclear, mitochondrial and peroxisomal proteins.
- Glycogen granules: Storage form of glucose (*replaced by starch in plants).
- Cytoskeleton: Interlinked network of three sets of protein filaments known as microtubules, microfilaments and intermediate filaments, which are collectively involved in the regulation of cell division, cell movement, cell shape and intracellular transport.

Basic Bioscience Laboratory Techniques: A Pocket Guide, First Edition.
Philip L.R. Bonner and Alan J. Hargreaves.
© 2011 John Wiley & Sons, Ltd. Published 2011 by John Wiley & Sons, Ltd.

Figure 2.1 *Diagram of a typical animal cell. Adapted from Nelson and Cox Lehninger: Principles of Biochemistry 5th Ed 9, WH Freeman.*

- Centrioles: a pair of these forms the poles of the mitotic spindle during cell division.
- Cytosol: A mixture of soluble proteins that carry out a variety of structural and enzymatic roles within the cell (e.g. enzymes involved in glycolysis).

In addition to the above structures, a number of additional features are found in plant cells, including:

- A large central vacuole: Uptake of water, storage of materials and degradation of macromolecules by acid hydrolases.
- Chloroplasts: Photosynthesis. Semi-autonomous organelles that synthesize some of their own proteins. Triple membrane (inner, outer and thylakoid) structure.

- Cellulose cell wall: Structural rigidity.
- Starch grains: storage form of glucose.

Ultrastructural detail of all the above macromolecular structures can be seen using transmission electron microscopy. Although structural detail of macromolecules cannot be visualized directly in the light microscope, it is possible to demonstrate the location and distribution of many macromolecules within cells and tissues with the aid of histochemical staining (see later). In combination with biochemical sub-fractionation techniques, microscopy has played a central role in the study of cell structure and organelle function.

Student exercise

- Draw a labelled diagram of a typical animal cell, indicating the main structural components.
- List any additional structures that are absent from animal cells but present in (a) plant cells and (b) yeasts.
- Indicate which of the structures found in eukaryotic cells are absent from prokaryotes.

2.2 Microscopes – General Principles

All microscopes consist of a coordinated system of lenses providing a magnified image of the specimen. They are invaluable tools in the study of the structure, organization and migration of cultured animal and bacterial cells. Microscopy can also be used to determine tissue organization and the viability, distribution and function of cells within tissues as well as the identification, organization and quantification of organelles and macromolecules within cells and tissues. Microscopy therefore has applications in all of the bioscience related subjects including microbiology, cell biology, biochemistry, biomedical sciences, forensic science, pathology, toxicology, plant biology, and so on and is particularly useful at providing supporting data for molecular studies of cellular components.

The type of microscope needed for a particular task depends on the level of detail required. The main differences between the different types of microscopes are the wavelength of electromagnetic radiation (see Chapter 3) used to obtain the image, the nature and arrangement of the lenses and the methods used to view the image. In this chapter we will look at the principles of light and electron microscopy, but first of all we will consider some general principles applicable to all types of microscopy.

2.3 Principles of Image Formation

There are three basic principles of image formation: resolving power, magnification and image formation.

i. Resolving power is the minimum distance apart that two points can be distinguished as two separate objects. The resolving power of the unaided eye is approximately 0.2 mm, while that of the light microscope is approximately 200 nm, whereas the transmission electron microscope can achieve a resolution of 0.1 nm.

ii. *Magnification* A common principle in all systems is that the objective lens forms an intermediary image, which becomes the object of the projector lens, which then produces the final image. For example, in a basic laboratory based light microscope the projector lens is the eyepiece, which projects an image on to the retina, a camera or a detector that provides an image suitable for viewing (e.g. on a monitor).

iii. *Contrast:* The level of contrast determines the visibility of the specimen caused by relative differences in light (or electron beam) intensity as it emerges from different areas of the specimen. Contrast can be enhanced by the use of chemical stains or a range of contrast enhancement techniques which will be discussed later. Specimens can contain amplitude objects or phase objects (Figure 2.2). Amplitude objects decrease intensity by scattering or absorbing light, whereas phase objects induce phase changes in the light passing through them which must be converted to intensity changes in order to see enhanced contrast. Intensity changes are then observable by eye.

2.3.1 Units of Measurement

The units of measurement you are most likely to come across in microscopy and their symbols are the metre (m), the centimetre (cm), the *millimetre* (mm), the *micrometre* (μm), the *nanometre* (nm) and the angstrom (Å), with the most common ones shown in italics (Table 2.1). The level of detail discernible by different microscope techniques is summarized in Table 2.2.

Student exercise

Memorize the units of measurement commonly used in microscopy and their interrelationships with each other.

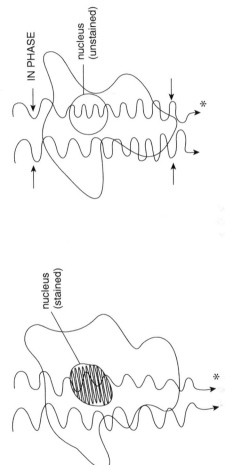

Figure 2.2 Schematic representation of amplitude and phase objects in light microscopy.

31

Table 2.1 *Conversion chart for units of measurement (distance) in microscopy*

Unit	Equivalent in other units					
	m	cm	mm	μm	nm	Å
m	—	10^2	10^3	10^6	10^9	10^{10}
cm	10^{-2}	—	10	10^4	10^7	10^8
mm	10^{-3}	10^{-1}	—	10^3	10^6	10^7
μm	10^{-6}	10^{-4}	10^{-3}	—	10^3	10^4
nm	10^{-9}	10^{-7}	10^{-6}	10^{-3}	—	10
Å	10^{-10}	10^{-8}	10^{-7}	10^{-4}	10^{-1}	—

2.4 Light Microscopy

2.4.1 Bright Field Microscopy

The earliest microscopes had a single convex lens and are referred to as simple microscopes. Modern light microscopes are called compound microscopes because they use a combination of biconvex objective and eyepiece lenses to provide a magnified image.

Table 2.2 *Relationship between object size ranges, microscopic technique required for detection and observation*

Observable detail	Size range
Atoms[*]	0.1–1 nm[*]
Proteins[**]	1–10 nm[**]
Macromolecular structures (e.g. cytoskeleton), ribosomes, small membrane vesicles	10–100 nm[**]
Mycoplasmas and small organelles (e.g. mitochondria)	0.1–1 μm[**]
Bacterial cells and larger organelles (e.g. chloroplasts and nuclei)	1–10 μm[***]
Eukaryotic cells	10–100 μm[***]
Tissue organization	100 μm–1 mm[***]

Key: [*]Observable only by electron microscopy
[**]Observable by electron microscopy; detectable by light microscopy but only when using histochemical staining techniques
[***]Observable by all techniques with appropriate chemical stains or optical enhancement methods

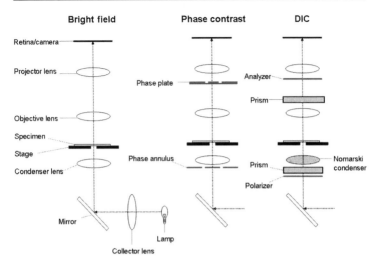

Figure 2.3 *Lens arrangements in bright field, phase contrast and DIC microscopy.*

The first compound microscope was developed by Janssen in 1595 with the capability to magnify an object from 10 to 30 times. The first observations at cellular level were made by Robert Hooke around 1665 who coined the term 'cell' for the cell-like units observed in cork tissue. However, the first observations of *living* cells (including bacteria, protists, algae, sperm cells and ciliates) were published by Antonie Van Leeuwenhoek between 1673 and 1700 using a simple microscope capable of magnifications up to 200 times.

Since that time, major improvements to the compound microscope have lead to superior magnification and image quality compared to the simple microscope. A typical lens arrangement for a compound microscope is shown in Figure 2.3.

2.4.1.1 Specimen Preparation

If a non-living specimen is to be observed, it is typically fixed using a solution of formalin (4% w/v p-formaldehyde), after which it is dehydrated and embedded in paraffin wax or a hard epoxy resin. Other methods of fixation may be used in specific circumstances. For example, cultured cell monolayers can be fixed at $-20\,°C$ with 90% (v/v) methanol and blood smears or bacteria are sometimes fixed to a slide by gentle heating above a Bunsen flame, after which they can be stained directly.

If tissue sections are being stained, the embedding process allows the sample to be sliced in to thin sections of a few microns in thickness, using a microtome. The sections are carefully transferred on to a glass slide before being incubated with chemical stains to enhance contrast on viewing in the light microscope. Some chemical stains that are commonly used in light microscopy are described in Table 2.3. In the case of unfixed or living specimens, samples are visualized directly in the light microscope taking

Table 2.3 *Examples of chemical stains and their uses*

Name (other names)	Examples	Staining pattern
Basic (cationic) dyes	Haematoxylin Crystal violet Giemsa Safranin	Stain basophilic structures blue or purple (e.g. heterochromatin). Commonly used for nuclear staining.
Acidic (anionic) dyes	Eosin Acid fuschin Fast green	Stain structures that contain positive charge (e.g. proteins, cytoplasm).
Protein binding dyes	Coomassie brilliant blue	Used to stain proteins in electrophoresis gels and thus stain the whole cell blue.
Metallic stains	Silver nitrate	Stains neurofibrils in neurones giving better contrast than H & E
Lipid stains	Sudan black	Stains phospholipids and other lipids. Can be used to identify granules in neutrophils on blood and bone marrow smears.
General stains	Toluidine blue	General cell stain used in plastic embedded sections. Some formulations used to stain carbohydrate polymers in cell walls and cartilage
Immunohistochemical stains	Horseradish peroxidase-conjugated antibodies	Specific antigen localization following incubation with chromogenic substrate that produces coloured insoluble deposits around antigen location

advantage of any natural pigments (e.g. chlorophyll) or using enhanced contrast optics (discussed later).

Many staining procedures involve a combination of two or more dyes, one often acting as a counter-stain. Specimen slides are often placed on slide racks or dipped in slide staining jars during the various staining steps. Two very commonly used staining protocols are outlined below.

2.4.1.2   Gram Stain for Bacteria

Method:

1. A heat-fixed smear of bacterial cells is flooded with crystal violet for 30 seconds before a quick rinse with water.
2. It is then flooded with Gram's iodine for 60 seconds then rinsed with water.
3. Next it is decolourized with 95% (v/v) ethanol then water-rinsed again.
4. The specimen is then counter-stained with safranin for 60 seconds, before being rinsed with water, carefully blotted dry prior to examination with the aid of a light microscope using an oil immersion lens.

Interpretation:

Gram-positive bacteria stain blue/violet because they retain the crystal violet after washing with 95% ethanol, as they have a much thicker cell wall than Gram-negative bacteria. The latter lose the dye during the ethanol wash and are eventually stained pink by the safranin counter stain, which is overwhelmed by the darker crystal violet stain in Gram-positive cells. Apart from revealing the presence of Gram-positive (purple), Gram-negative (pink) or both types of cells, the shape of cells present (Bacilli, cocci, etc.) can also be distinguished in the light microscope. Care needs to be taken not to over-stain or decolourize the specimen for too long, as this could lead to false-positive or false-negative results, respectively. Gram staining is almost always the first step performed in identification of bacteria from a variety of isolates taken from patients, industry, food, food preparation areas, and the environment.

2.4.1.3   Haematoxylin/Eosin Staining of Paraffin Wax Sections

Method:

1. Thin sections are deparaffinized (e.g. by gentle warming and a 3-minute wash in xylene).

2. They are then rehydrated by consecutive washes in absolute ethanol (3 × 3 min), 95% (v/v) ethanol (1 × 3 min) and 80% (v/v) ethanol (1 × 3 min) finishing with a 5-minute wash in distilled water.

3. Specimens are then stained for 3 minutes in haematoxylin, rinsed in deionized water, then incubated in tap water for 5 minutes to allow colour development.

4. Destaining is achieved by dipping quickly 8–12 times in acid ethanol, followed by two 1-minute rinses in tap water and a 2-minute rinse in deionized water.

5. Excess water is carefully blotted away from the slide before dipping in the eosin counter-staining solution for 30 seconds, followed by three 5-minute washes in 95% (v/v) ethanol and three 5-minute washes in absolute ethanol.

6. After three 15-minute washes in xylene, a drop of xylene-based slide mountant (e.g. Permount) is applied on to the cover-slip, avoiding bubbles.

7. The cover-slip is carefully placed on the slide such that the mountant spreads evenly between it and the glass slide. After drying, the specimen is ready to view in the light microscope.

Interpretation:

Nuclei stain blue/violet with haematoxylin and the cytoplasm stains pink with eosin in most eukaryotic cell types. Mammalian red blood cells can be distinguished in tissue sections by their ellipsoid shape, small size and pink colour (due to the absence of nuclei on maturation) their presence in a clear area (capillary lumen) surrounded by a an ordered cell layer (capillary endothelial cells). It is possible to observe tissue damage and tissue organization using this technique which allows the identification of different cell types (e.g. by their distinctive shape, size, or staining pattern. Haematoxylin and eosin (H & E) staining can also be utilized to estimate mitotic index (i.e. the percentage of cells showing condensed chromatin arrangements corresponding to chromosomes at various stages of cell division) and to identify potential tumour cell growths.

2.4.2 Phase Contrast Microscopy

This method utilizes contrast enhancement optics so that unstained fixed or living tissue can be observed. Bright field microscopy detects reduced amplitude of light to generate contrast, whereas phase contrast microscopy detects changes in refractive index. Conversion of a microscope from bright field to phase contrast can be achieved by the inclusion of a phase ring and a phase plate in the light path (see Figure 2.3). Diffracted light is spread evenly, whereas a hollow cone of undiffracted light is passed through the specimen. The phase optics allow in-phase and out of phase light to be recombined, resulting in enhanced contrast. A potential drawback with this method is the appearance of a bright ring particularly around small spherical objects, sometimes referred to as the 'halo effect'.

2.4.3 Differential Interference Contrast (DIC)

This technique uses special optics to detect changes induced in the orientation of plane polarized light by birefringent objects. DIC utilizes the interference between separated wave fronts, giving highest contrast where the differential is greatest. Conversion from bright field to DIC requires the use of a specialized 'Nomarski' condenser unit, and the addition of an upper prism, a secondary prism, an analyzer and a polarizing filter, as indicated in Figure 2.3. The images produced appear to be three-dimensional due to a shadow-cast effect of polarized light. DIC has several advantages over phase contrast, including the abolition of the halo effect and objective lenses with a shallower depth of field, enabling optical sectioning.

2.4.4 Fluorescence Microscopy

This specialized technique allows the detection of small molecules using fluorescent probes (e.g. antibodies labelled with fluorophores). Fluorophores are molecules that can be excited to a higher energy state by irradiation with an appropriate wavelength of light (excitation wavelength), subsequently releasing some of the absorbed energy in the form of fluorescence at a longer wavelength (the emission wavelength). Fluorophores can be covalently attached to a variety of probes to detect molecules that bind the probe. For example, staining of tissue sections with fluorescently-labelled antibodies allows the localization of anti-gens by immunofluorescence staining.

   Visualization of fluorescence (see Chapter 3) requires the use of a fluorescence microscope, which normally uses a mercury lamp and a set of wavelength filters and beam splitters to irradiate the specimen evenly through the objective lens (acting as the condenser lens) (Figure 2.4). Most of the light passes through the specimen but some is absorbed, releasing energy in the form of fluorescence at a longer wavelength, which passes back through the objective lens and straight through the beam splitter up to the eyepiece.

   A major problem with this approach can be interference from out-of-focus fluorescence, which gives a fuzzy background. This can be eliminated using a

Figure 2.4 *Lens arrangement and light path in epifluorescence microscopy.*

confocal laser scanning fluorescence microscope, which scans the specimen at different heights through a narrow plan of focus using precise laser beams to eliminate stray light interference.

2.4.5 How to Use a Light Microscope

Rather than getting a big enough magnification, the main challenges in light microscopy are obtaining sufficient contrast and resolution, locating the focal plane and recognizing the structures being viewed. The microscope you are most likely to come across in your studies is the bright field microscope.

2.4.5.1 *General Features of Microscopes*

In a conventional laboratory bright field microscope, the objective and eyepiece lenses are located above the specimen stage and the light source below it, as indicated in Figure 2.5. In the case of the inverted microscope, typically used in cell culture work, the objective lenses and light source are below the sample stage but the eyepiece lenses remain above it. A good microscope has a built in

Figure 2.5 *Diagram of a typical light microscope.*

illuminator, an adjustable condenser with an aperture diaphragm (which can be reduced in diameter to increase contrast up to a point), a mechanical stage and a binocular arrangement of eyepiece lenses.

The purpose of the condenser is to focus an even beam of light on the specimen, which, after passing through the objective and eyepiece lenses, is significantly magnified by the time it reaches the eye. If you are doing bright field microscopy and there are adjustable settings on the condenser, make sure that it is set to bright field. Some condensers are fixed whereas others can be focused to adjust the quality of the light. If focusable, the best position for the condenser is as close as possible to the microscope stage. The condenser is usually fitted with an aperture diaphragm, which can be opened to increase brightness at the expense of contrast and vice versa. When closing the aperture to increase contrast, a point is reached where the image becomes distorted. If the condenser requires adjustment, you should follow instructions given in the class or in the instrument manual, as precise details may vary from one microscope to another.

2.4.5.2   *Focusing a Specimen*

- Place the specimen slide carefully on the microscope stage with the light source switched on. There may be some slide clips or a groove to hold the slide in position.
- Make sure that the cover-slip that covers the specimen is facing towards the objective lens, as high power lenses are unable to focus through the relatively thick glass slide. Place the sample in the light beam and bring it into focus using a low magnification objective lens (typically $\times 10$ or $\times 4$).
- This is relatively easy to do if the sample is sectioned and stained but if using microbial cells, it can be useful to look for the edges of smears, where cell density is usually higher.
- In a good quality microscope the stage can be moved mechanically, enabling precise movements of the slide without touching it.
- Start with the specimen out of focus so that the stage and objective need to be moved closer together. The first surface coming in to focus will be the top of the cover slip, although this is not always applied over smeared samples, then the lower surface of the cover slip which is the plane of focus for the sample.
- Once the sample is in focus at low magnification, it is now possible to swing the higher magnification lens into position and refocus to obtain the higher magnification image. The sample may need to be re-focused and re-centred although with good quality lens systems the specimen should come into focus as the lenses are changed. It will also be necessary to adjust the light as higher magnification lenses to cover a smaller field of view and therefore less light passes into the lens at a given light intensity. Thus, with a low power lens the illumination intensity can be decreased and vice versa for high power objectives.
- The contrast and light intensity can then be adjusted and the sample position moved to obtain an optimal image. If oil immersion is required (e.g. with high power lenses $\times 100$), a drop of oil is placed on top of the cover-slip area above the sample before bringing it into contact with the oil immersion objective lens while focusing. The oil has the same refractive index as the glass slide, which enhances resolution. The oil immersion lens is essential for microbiological samples, especially bacteria.
- Remember that there are usually crude (outer) and fine (inner) focus controls, the latter being used at higher magnifications. However, the high power lenses

need to be much closer to the specimen; care is therefore needed to prevent contact with the cover-slip, which could jam the slide in position.

- Remember also that the cover-slip should be fixed in position, particularly if the sample is covered by a layer of liquid (e.g. in fluorescence microscopy).

HOW TO LOOK AFTER A MICROSCOPE:

All of the parts of a good quality microscope are expensive, so care is needed when handling and using a microscope. In particular:

- When moving a microscope, always hold it firmly by the stand.
- When unplugging the light source, always hold the plug (not the cable).
- As bulbs are expensive and have a limited life span, the illuminator should be turned off as soon as you have finished.
- Never use immersion oil on a dry lens, as the image will be distorted.
- Always clean the lenses (gently with proper lens tissue only!) and the stage before leaving the microscope.
- An appropriate lens cleaning fluid may be applied to remove dried material but some organic solvents may damage lens surfaces
- Place the designated dust cover on the microscope when it is not in use.
- Try to focus gradually and not too quickly, without forcing the focusing controls. If you encounter resistance, you have probably gone too far past the specimen and need to start again.
- If moving between samples under two different cover-slips on the same microscope slide using oil immersion lenses, avoid the risk of scratching the lens by lowering the microscope stage before moving the slide to the new position and refocusing.
- Never remove the lenses from the microscope unless you know what you are doing or are instructed to do so.

2.4.6 Data Analysis and Presentation

2.4.6.1 *Drawing and Labelling a Microscope Image*

- Remember to give each image a clear descriptive title.
- If drawing the image, draw representative areas of cells rather than every detail. Label the major structures or cell types visible in the field of view.
- If drawn by hand, indicate the magnification at which the specimen was viewed or, if using a camera image, include the final magnification (including image enlargement) or a calibrated scale bar to give a clear idea of dimensions and distances.

Magnification can be determined by multiplying the magnifying powers of the eyepiece (ocular) and the objective lens used to view the specimen (see worked example 2.1).

Worked Example 2.1: Calculation of image magnification

The magnification observed using a ×10 eyepiece lens and a ×40 objective lens would be:

$$10 \times 40 = 400\text{-fold magnification}$$

Note that, if making prints from a negative, the enlargement factor of the print from the negative also needs to be taken into account when calculating the final magnification. Many microscope cameras are digital and the camera software allows the insertion of a scale bar on to the image, which would adjust according to the final image size.

2.4.6.2 Making Size Measurements Using the Light Microscope

Microscopes may be fitted with a scale (reticule) in one eyepiece lens, which can be turned and aligned with any focused object to obtain an estimate of its size. The size of the object is initially noted as the number of ocular divisions it spans. This number is then multiplied by a conversion factor which is different for each objective lens and obtained by calibrating it against another micrometer scale etched onto on a glass slide and brought into focus on the microscope stage (see worked example 2.2).

A typical eyepiece reticule has a scale of 50–100 equal divisions, which can be brought into focus by adjusting the eyepiece focus if necessary. A typical stage micrometer scale is 2 mm long and it has areas with specific divisions (e.g. 0.01 mm or 10 μm). For calibration, these two scales are aligned as shown in Figure 2.6.

Figure 2.6 *Eyepiece reticule calibration. Adapted from Nelson and Cox Lehninger: Principles of Biochemistry 5th Ed 9, WH Freeman.*

42

Worked Example 2.2: Calibration of lenses with an eyepiece reticule and stage graticule

- Let's suppose that the alignment shown in Figure 2.6 was carried out using a ×100 objective lens.
- If each stage graticule division corresponds to 10 μm then, at this magnification, one eyepiece unit would be equivalent to 1 μm.
- At lower magnification this value would increase as the eyepiece reticule scale stays the same while the objective lens magnification and, therefore, the stage graticule image size decrease.
- Thus, with ×40 and ×20 objective lenses on the same microscope, one eyepiece reticule unit would be 2.5 μm and 5 μm, respectively.

When making single estimations of length or distance, you should bear in mind the fact that the actual limits of resolution of a dry objective lens (up to ×40) and an oil immersion lens (×100) on a basic laboratory instrument is unlikely to be better than 1 μm or 0.5 μm, respectively. Values should therefore be rounded up or down to the nearest whole number. For example, if you calculate the size of an object to be 4.55 μm, you should round up this value to 5 μm (*not* 5.0 μm, which suggests an accuracy of 0.1 μm). Of course, accuracy could be improved by making several measurements of independent objects and calculating the mean ± SD (see Chapter 4), in which case the use of precise estimates to one or two decimal places would be more meaningful.

2.4.6.3 Quantification of Specific Features in Microscope Specimens

In a laboratory class you may only be required to make a few observations or measurements (e.g. number of mitotic cells, number of red blood cells per unit area, etc.) on one or two specimens. However, with a small number of microscope fields you cannot be sure that you are viewing a representative area of tissue. In fact, to produce publishable research data, multiple fields in several independent specimens would be analyzed in order to take experimental variation into account. Within the time constraints of a laboratory class, comparison of your observations with those of other groups will give you a better idea of the reliability of your observations. Obviously, major differences in measurements between individuals may be due to inconsistencies in the method of measurement especially if the same specimens were analyzed. Alternatively, if all of the samples are different sections, differences could also be due to the orientation of the sections cut.

2.4.6.4 *Application of Microscopy in Cell Counting*

A simple example of quantification using the light microscope is the use of a haemocytometer chamber to count cells in a cell suspension. In this instance, the cells are suspended in an isotonic medium and a small volume of the suspension applied under the cover-slip of a haemocytometer chamber. On focusing the microscope, an image like the one in Figure 2.7 will appear. In the actual focused slide, triple lines (depicted by thicker lines in the diagram) delineate the major areas from which cells are to be counted.

Cultured cells are normally suspended in a medium, diluted in a buffer or a medium, before 10 µl are applied to the chamber beneath the cover-slip. If the cell number is overwhelming, the suspension can be diluted (1/10 to 1/200) and recounted. Cell density (i.e. the number of cells/ml) can be estimated in one of the following ways (see worked example 2.3).

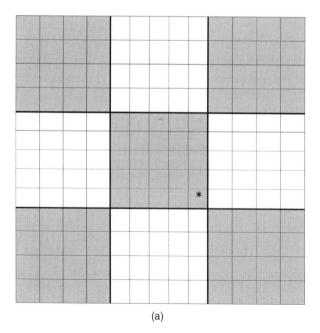

(a)

Figure 2.7 *(a and b) Counting cells in a haemocytometer chamber.*

(b)

Figure 2.7 (Continued) Shown is a schematic representation of a haemocytometer chamber grid viewed in the light microscope. In (a) each large shaded counting area covers 1 cm², having a volume of 0.1 μL. There are two different sizes of small squares in the counting areas. Those within the central large area have the dimensions 0.2 mm × 0.2 mm and there are 25 per large area. Those in the outer 4 large areas are 0.25 mm × 0.25 mm and there are 16 per large area. The dark grid lines in the diagram usually appear as a triplet of parallel grid lines in the microscope field of view. (b) The enlargement of the single small square marked with an asterisk shows that cells appear as spheres. Only cells within the boundaries of the large squares bordered by these triple lines should be counted. Cells that are outside or that overlap with the middle line of the triple line boundaries (shaded) should be excluded from the counts. Adapted from Nelson and Cox Lehninger: Principles of Biochemistry 5th Ed 9, WH Freeman.

a. For most mammalian cells in suspension (e.g. cultured cells), count all of the cells lying within five large areas, then estimate the cell density (cells/ml suspension) by applying the formula:

Mean count per large area × dilution factor (typically 10 or 20) × 10^4 (volume factor)

b. For leucocytes, count the cells in the four large corner squares, then calculate the number of cells/ml from the formula:

Average cell counts per large corner square × dilution factor (typically 20) × 10^4

c. For a suspension of red cells, count the erythrocytes in five *small* squares in the middle large area, estimating the cell density (cells/ml) by the formula:

(Total counts from 5 squares × 5) × dilution factor (typically 200) × 10^4

Worked Example 2.3: Estimation of cell density from counts in a haemocytometer chamber

Estimate the cell density of a suspension of white blood cells which, when diluted 10-fold gives counts of 55, 67, 49 and 70 cells in the four outer large counting areas of a haemocytometer grid.

- First of all, work out the average cell count per large counting area, as follows:
  $(55 + 67 + 49 + 70)/4 = 60.25 = 60$ (rounded down to nearest whole number)
- Incorporating this value into the equation described above for leucocytes:
  Cell density $= 60 \times 10 \times 10^4 = 6 \times 10^6$ cells/ml

2.5 Electron Microscopy

As indicated in the general microscopy section, the electron microscope offers much better resolution than the light microscope and a much higher magnification range. The two main types are known as transmission (TEM) and scanning (SEM) electron microscopes, which use electron beams differently to obtain images with different characteristics, as indicated in Table 2.4.

- When viewing images of thin sections in the TEM, you should bear in mind that the profiles of larger organelles observed will be affected by the thickness of the section, the location of specific organelles and by the angle of the section relative to the organelle profile.
- Thus a sausage-shaped mitochondrion might appear spherical if caught in cross-section, more sausage shaped in a symmetrical longitudinal section or something in between. Additionally, two spherical profiles appearing side by side in a thin section could be two separate mitochondria or the same long mitochondrion moving in and out of the plane of the section.
- Changes in cell structure can be estimated using morphometric measurements to count the number of profiles corresponding to particular organelles per unit area. Bearing in mind what has been said, it is unlikely that a single section can be representative of the whole cell.
- Therefore, many images should be analysed from different parts of the specimen and from several specimens under each experimental condition in order to carry out statistically valid estimations of organelle distribution.

Table 2.4 Properties of electron microscopes and sample preparation methods

Property	TEM	SEM
Maximum magnification	1 000 000	300 000
Typical limit of resolution	0.2 nm	1.5 nm
Type of detail discernible	Molecular and macromolecular structure (e.g. molecules, membranes, cytoskeletal elements, chromatin, viruses, etc.). Cross or longitudinal sections.	General and fine surface details of whole cells, tissues and small organisms. Surface detail in 3D image.
Electron beam	Transmitted through stained specimen	Reflected by specimen coated in gold
Sample preparation	Classical approach: Specimen is fixed, dehydrated, embedded in resin, sectioned, stained with lead citrate and uranyl acetate, dried and then viewed by TEM.

Freeze fracture: Tissue is rapid frozen at $-140\,^{\circ}$C, freeze-dried, coated with platinum, after which the sample is digested, and the replica viewed by TEM. Direct negative staining: Suspension of fixed or unfixed macromolecules (DNA, viruses, and cytoskeletal networks) is adsorbed on to a plastic coated EM grid and stained briefly with uranyl acetate, dried and viewed by TEM. | Specimens are typically fixed, dried, coated (e.g. with gold), then scanned with a very narrow beam of electrons. Secondary electrons emitted by specimen are detected in a cathode ray tube, creating an image of the specimen surface. |

- The most common approach to quantify changes in organelle composition or volume relative to total cell volume is to superimpose a grid template with grid points at a fixed distance apart (e.g. 1 cm). The organelle or structure in contact with each point of the grid is recorded and can be estimated as a percentage of the total volume (grid points).
- Clearly this is only a very rough estimate and would be subject to the errors discussed previously for the angle and position of sections; reliability of this approach would also be affected by the accuracy of identification of the cell structure and the number of points on the grid (see worked example 2.4).
- Once again, measurements on multiple images for several sections would give statistically meaningful data as a single section may not necessarily be representative of the whole specimen.

Worked Example 2.4: morphometric analysis of electron micrographs

If a grid were superimposed on the image shown schematically in Figure 2.8, and coincidence of grid points (points of intersection of grid lines) with organelles or 'ground cytoplasm' recorded, it is possible to get an estimate of the abundance of each organelle as % total cell volume as follows:

Total number of grid points = 48. Number of 'hits' is expressed as a % of total grid points.

Figure 2.8 *Electron microscope grid morphometry.*

Mitochondria	=	$(3/48) \times 100 = 6.2\%$
Nucleus	=	$(12/48) \times 100 = 25.0\%$
Golgi apparatus	=	$(2/48) \times 100 = 4.2\%$
RER	=	$(2/48) \times 100 = 4.2\%$
SER	=	$(3/48) \times 100 = 6.3\%$
Free polyribosomes	=	$(1/48) \times 100 = 2.1\%$
Ground cytoplasm	=	$(25/48) \times 100 = 52.0\%$

It is possible to estimate the actual size of cell structures from a photographic image provided that the final magnification is known (see worked example 2.5). In this case the measurement (e.g. in mm) is divided by the magnification factor, which should include the factor for photographic enlargement. The value is then converted into appropriate units.

Worked Example 2.5: calculation of organelle sizes from electron micrographs

In an image with a final magnification of $\times 80\,000$, a structure identified as a mitochondrion was found to be 20 mm in length and 13 mm in width. What are its actual dimensions?

First of all, divide the measured size in the image by the magnification factor to obtain the size in mm. Then, referring to Table 2.1, convert the value into appropriate units.

a. Mitochondrion length $= 20 \div 80\,000 = 0.00025$ mm
These are not appropriate units as organelle size is normally expressed in µm or nm.
Length in µm $= 0.00025 \div 1000 = 0.25$ µm
Length in nm $= 0.00025 \div 10^6 = 250$ nm
b. Mitochondrion width $= 13 \div 80000 = 0.0001626$ mm
Length in µm $= 0.0001626 \div 1000 = 0.16$ µm (to 2 decimal places)
Width in nm $= 0.0001626 \div 10^6 = 163$ nm (rounded up to nearest whole number)

2.6 Summary

- Light microscopy has a limit of resolution of approximately 200 nm and a maximum magnification of $\times 1500$, allowing visualization of structures down to the size of mitochondria.

- Electron microscopy can achieve a limit of resolution down to 0.2 nm (TEM) and a maximum magnification of $\times 10^6$, allowing visualization of ultrastructural detail of the cell.
- Bright field microscopy is used to determine cellular organization at the tissue level.
- Contrast enhancement in light microscopy is achieved by the use of chemical stains such as haematoxylin/eosin or, in the case of unstained specimens, by the use of contrast enhancement optics in phase contrast or differential interference contrast microscopy.
- The tissue and intracellular distribution of specific molecular components can be observed using antibody staining techniques, although the actual visualization of the molecules at an ultrastructural level requires the use of transmission electron microscopy.
- Contrast in TEM is enhanced by staining with electron dense heavy metals such as uranyl acetate.
- Interpretation of electron and light micrograph images is aided by developing knowledge of (a) general tissue organization and (b) the appearance and location of organelles within cells.
- When interpreting images, remember that the appearance of a specimen in a thin section is partly dependent on the angle of cut.

3

SPECTROPHOTOMETRY

3.1 Introduction

A popular instrument used by students on biological sciences courses in higher and further education is the spectrophotometer. Proteins, peptides, amino acids, nucleic acids, nucleotides and other cellular metabolites can be measured by the application of spectrophotometry. In this section we will focus on the wavelengths of the electromagnetic spectrum (200–800 nm) covering the wavelengths of ultraviolet light (200–350 nm) and visible light (350–800 nm).

Many laboratory procedures require the use of a spectrophotometer (a machine that measures the transmission and absorbance of light) and modern spectrophotometers are easy to use. However, to fully understand the results generated and to avoid any errors, it is advisable to have a basic understanding of the machines and of the physical laws involved.

3.2 The Electromagnetic Spectrum

The electromagnetic spectrum (see Figure 3.1) is a collective term given to a broad band of radiations, ranging from the longer wavelengths (see Figure 3.2) of radio waves (wavelengths ≥ 1.0 cm) to the shorter more energetic wavelengths of the gamma rays (wavelengths ≤ 1.0 pm). Photons (hv) are the basic unit of the electromagnetic spectrum at every wavelength and they have the properties of both a wave and a particle (wave-particle duality).

These properties include:

- They have zero mass.
- In free space they move at the speed of light ($c = 3.0 \times 10^8$ ms^{-1}).
- They are discrete packages of energy, which is related to their wavelength (shorter wavelengths have more energy; see worked example 3.1).
- They can be absorbed (destroyed) and created (emitted).[1]
- They can have particle like interactions (i.e. collisions) with electrons and other atomic particles.

Basic Bioscience Laboratory Techniques: A Pocket Guide, First Edition.
Philip L.R. Bonner and Alan J. Hargreaves.
© 2011 John Wiley & Sons, Ltd. Published 2011 by John Wiley & Sons, Ltd.

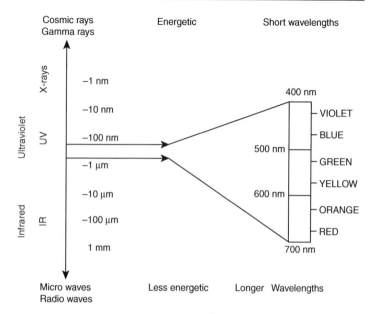

Figure 3.1 *The electromagnetic spectrum.*

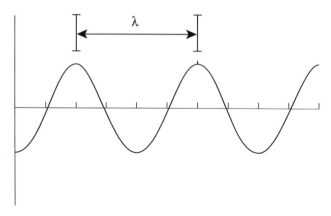

Figure 3.2 *Wavelength and frequency.*

52

Key to Figure 3.2

λ: (lambda) wavelength, distance between wave crests
v: (nu) frequency, number of wave crests that pass a point in a given amount of time.

Wavelength and frequency are related:

$C = \lambda v$ (where c = the speed of light, 3.0×10^8 m s^{-1})
$v = c/\lambda$

Worked Example 3.1: Different electromagnetic wavelengths have different energy levels

Max Planck showed that the energy available at a given wavelength is given by $E = hv$

where

h = Planck constant (6.63×10^{-34} Js)
v = wave frequency

The wave frequency $\quad v = c/\lambda$

Where

c = speed of light (3×10^8 ms^{-1})
λ = wavelength (usually in nm)

- *The energy of a photon at 300 nm*

$$v = c/\lambda$$

$$\frac{3 \times 10^8 \text{ ms}^{-1}}{300 \times 10^{-9} \text{ m}} = 1 \times 10^{15} \text{ Hertz}$$

$$E = hv$$

$(6.63 \times 10^{-34}) \times (1 \times 10^{15}) = 6.63 \times 10^{-19}$ Joules per photon at 300 nm

- *The energy of a photon at 600 nm*

$$v = c/\lambda$$

$$\frac{3 \times 10^8 \text{ ms}^{-1}}{600 \times 10^{-9} \text{ m}} = 0.5 \times 10^{15} \text{ Hertz}$$

$$E = hv$$

$6.63 \times 10^{-34} \times 0.5 \times 10^{15} = 3.32 \times 10^{-19}$ Joules per photon at 600 nm.
The energy of a photon with a wavelength of 300 nm is double the energy of a photon with a wavelength of 600 nm.

- *Conclusion: Photons with shorter wavelengths have more energy*

3.3 The Absorbance of Light

Every atom of every molecule has a defined structure with negatively charged electrons rotating in orbitals around the positively charged nucleus of the atom. Each electron orbital has an energy requirement for occupancy, that is if an electron receives additional energy it can no longer rotate in its current orbital and is promoted to a vacant orbital with a higher energy requirement. Photons radiated by the sun across the electromagnetic spectrum have wave-like properties (discrete aliquots of energy) and particle-like properties (the ability to interact with other atomic particles) described as wave-particle duality.

When the wide range of photons from the electromagnetic spectrum encounter the atoms of molecules, some of the photons will have the correct amount of energy to promote electrons to higher energy orbitals. If the energy requirement of the electron is matched by the energy in the photon, the energy is transferred to the electron. The electron now has too much energy to remain in the ground state orbital and is promoted to a higher energy orbital (see Figure 3.3). The photon's energy is taken up by the electron, which stays in the higher energy orbital, rotating for a brief period of time (1×10^{-15} sec) using up the energy supplied by the photon.

The photon's energy has been *absorbed* in its entirety and the photon at that wavelength will no longer exist. Having used up the energy supplied by the photon, the electron returns to its original orbital in the ground state and the whole process can be repeated by the absorption of another photon. Other photons at different wavelengths will not have the exact energy match for the electron and will pass through the atoms of the material.

3.3.1 Fluorescence

In general, molecules that fluoresce are organic molecules that have associated ring structures and they have both excitation and emission spectra (determined by a scan; see Figure 3.4). In fluorescence, when a photon is absorbed at the excitation wavelength, the excited electron is elevated to higher energy levels. In the higher energy orbital, the electron exhausts some of the received energy; as a result it no longer has sufficient energy to remain at the higher level and must return to the ground state. However, to be able to re enter the ground state the electron must shed the remaining energy by re-emitting a photon ('fluorescence') at a longer wavelength[2] (i.e. with less energy). Fluorescence spectroscopy can be used to increase the sensitivity of assays and of the detection of biological components.

54

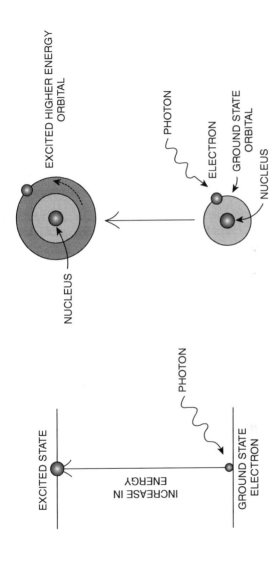

Figure 3.3 *The absorption of a photon.*

Figure 3.4 (a) A scan between 200–400 nm of 100 mM NAD$^+$ using a Beckman 530 spectrophotometer. (b) A scan between 200–400 nm of 100 mM NADH using a Beckman 530 spectrophotometer (ABS = absorbance).

3.3.2 Luminescence

The absorbance of light and fluorescence by molecules depends upon receiving energy from the electromagnetic spectrum. In contrast, luminescence is the generation of wavelengths of light as a result of a chemical reaction. Chemiluminescence occurs as a result of a chemical reaction and bioluminescence occurs as a result of an enzyme catalyzed reaction. Bioluminescent assays are increasingly popular because of their high sensitivity. Luciferase from fireflies can be used to detect 1×10^{-15} M ATP and bacterial luciferase can be used to detect 1×10^{-12} M NADH, NADPH and FMNH$_2$.

3.4 Absorption Spectrophotometry

The visible wavelengths of the electromagnetic spectrum are composed of the colours of the rainbow and the colour of any material is a result of the absorbance of different wavelengths, for example plants appear green in colour

(500–600 nm) because they contain pigments which absorb the blue wavelengths of light between 400–500 nm and the red light wavelengths of between 600–700 nm. The wavelengths that are not absorbed are the wavelengths we can see. In the case of plants, the remaining wavelengths of light (500–600 nm) appear green. The colour of any compound depends upon its structure.

KEY POINTS TO REMEMBER 3.1

The colour of material or a solution is due to the wavelengths of light which have not been absorbed for example a solution of copper sulphate is blue because, as white light passes through the liquid, the wavelengths of light which make up the colours red, green and yellow are absorbed.

The wavelengths of light that pass through a solution can be measured by a photosensitive device and this forms the basis of spectrophotometry. Spectrophotometers (see Section 3.7) can measure the absorbance of a solution at all the different wavelengths of UV/visible light and plot these measurements to provide a 'scan'. The scan (see Figure 3.4a and b) shows that compounds can absorb photons of light across a wide range of the UV/visible part of the EM spectrum. However, there is usually a wavelength at which the compound will absorb the most photons. This maximally absorbing wavelength is called the lambda max (λmax e.g. λ_{410nm}) and this is characteristic of the compound absorbing the light.

KEY POINTS TO REMEMBER 3.2

Changes in the structure of molecules can subtly alter their absorbance pattern, for example nicotinamide adenine dinucleotide (NAD^+) (Figure 3.4a) and reduced nicotinamide adenine dinucleotide (NADH) (Figure 3.4b) are vital cellular cofactors. Both molecules absorb light at 260 nm, but only the reduced cofactor (NADH) also absorbs light at 340 nm.

3.5 The Laws of the Absorbance of Light

The transmission of light is measured by the percentage of the incident light that passes through the sample:

$$\% \text{ Transmittance(T)} = (I/I_0) \times 100\%$$

where:

I = intensity of transmitted light and

I_0 = intensity of incident light. In this case intensity is the number of photons interacting second^{-1}

When % Transmittance (T) = 100% a solution is totally transparent and when % Transmittance (T) = 0% a solution is opaque.

If T = 40%, then 40% of the photons passing through the sample reach the detector and the other 60% are absorbed by the sample.

Absorbance (see Figure 3.5) is the $\log_{10}$ of the incident light (I_0) divided by the transmitted light (I).

$$A = \log_{10}(I_0/I) = \log_{10}(1/T)$$

Absorbance (a logarithmic value) has no units and takes values between zero (0) and infinity (∞)

- If $A = 0$ then no photons are absorbed.
- If $A = 1.00$ then 90% of the photons are absorbed; only 10% of the photons reach the detector %T = 10%.
- If $A = 2.00$ then 99% of the photons are absorbed; only 1% of the photons reach the detector %T = 1%.
- Most spectrophotometers will be able to display the absorbance of a sample and/or the transmitted light through the sample.

3.5.1 Attenuance of Light (Optical Density)

Absorbance and light scattering (attenuance or optical density) are different phenomena. The absorbance of light at a wavelength by a liquid is a measure of the photons captured by a transparent solution. When solutions containing cells, plastids and particles are placed in a spectrophotometer they may absorb light but they will additionally also deflect light from the detector resulting in a decreased signal. This reduction in measured light given by these cloudy solutions is called the attenuance of light. In microbiology the term optical density (OD) is used.

KEY POINTS TO REMEMBER 3.3

The absorbance of light is not the equivalent of the attenuance (optical density) of light. When the particulate containing solution is sufficiently diluted, attenuance will approximate to absorbance.

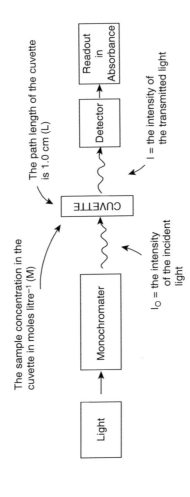

Figure 3.5 *Incident and transmitted light.*

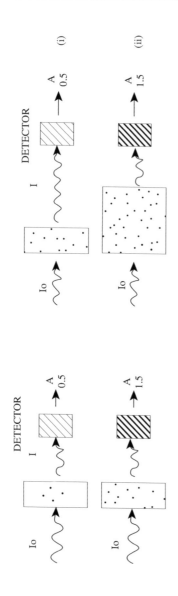

(a)

The concentration in cuvette (ii) is 3 times the concentration in cuvette (i). This is reflected in the increase in absorbance.

(b)

The concentration of the solution in b (i) is the same as in b(ii) but the path length in cuvette (ii) is 3 times longer than the path length in cuvette (i). This is reflected in the increase in absorbance.

Figure 3.6 *Beer Lamberts Law is related to (a) the concentration of the sample and (b) the path length that the light has to travel through.*

3.6 The Beer Lambert Law

The majority of scientists choose to report data in terms of absorbance rather than transmittance, because absorbance is related to the concentration of the solution. In a spectrophotometer the absorbance readout is a measure of the amount of photons absorbed by a solution. It follows that if there are more molecules in the solution (i.e. at a higher concentration) the absorbance will be greater (see Figure 3.6a). This relationship is linear, in that as the concentration of a solution increases so does the absorbance.

In a similar manner, the longer the path length that light has to travel through a solution before reaching the detector the greater the chance a photon will be absorbed. Therefore, absorbance is also related to the path length the light has to travel through the solution (see Figure 3.6b).

The Beer Lambert law states that the absorbance of a solution is a function of three factors:

- the path length (L) the light has to pass through the cuvette holding the solution (this has been standardized in most laboratory instruments to be 1.0 cm).
- the concentration (C) of the reagent in molarity (M)
- the absorptivity ($\in$), which is characteristic of the analyte in solution ($\in$ is called molar absorptivity ($M^{-1}\,cm^{-1}$) if the concentration is measured in molar (M); see Chapter 1).[3]

$$\text{The Beer Lamberts Law} \qquad \text{Absorbance}(A) = L \in C$$

where:

$\in$ = molar absorptivity coefficient ($M^{-1}\,cm^{-1}$)
C = molar concentration of solute (M)
L = the path length of sample cell/cuvette (cm).

It is the absorbance that displays a simple dependence on the concentration of a solution and the path length the light has to travel through (Beer Lambert Law). This relationship allows scientists to convert an absorbance value from a spectrophotometer into a concentration.

KEY POINTS TO REMEMBER 3.4

The molar absorptivity coefficient ($\in$) is valid only for the specified wavelength. The value for ($\in$) for many different compounds may be obtained from reference books and this represents the absorbance of a 1.0 M solution at a *defined* wavelength in a 1.0 cm cell/cuvette path length at 25 °C. For example, the molar absorptivity coefficient ($\in$) of NADH at 340 nm is **6.22×10^3** ($M^{-1}\,cm^{-1}$) at 25 °C.

Spectrophotometers measure the absorbance of solutions between 0–3.0 (see Section 3.5). This means that a 1.0 M solution of NADH will need to be diluted to bring it onto the scale of the machine.

- A 1.0 M solution of NADH at 340 nm would have an absorbance of 6220
- 1 : 10 dilution results in a 100 mM solution of NADH with an absorbance of 622
- 1 : 100 dilution results in a 10 mM solution of NADH with an absorbance of 62
- 1 : 1000 dilution results in a 1 mM solution of NADH with an absorbance of 6.2
- 1 : 10 000 dilution results in a 100 μM solution of NADH with an absorbance of 0.62.

A spectrophotometer will be able to provide an absorbance reading when the concentration of NADH is between 2–400 μM.

3.6.1 Limitations of the Beer Lambert Law

The Beer Lambert Law is a linear relationship between the absorbance and the concentration up to 1.0 M. In reality, a plot of concentration against absorbance may deviate from linearity at high concentration values forming a plateau, possibly due to interactions between molecules. The Beer Lambert Law will not apply in the plateau area of the plot and readings should be relied upon only in the *linear* part of the graph.

Worked Example 3.2

The molar absorptivity coefficient ($\in$) of NADH at 340 nm is **6.22×10^3** ($M^{-1} cm^{-1}$) at 25 °C and the path length of the cell is **1.0** cm. The relative molecular mass (Mr) of NADH is 663. (Please note that the wavelength number and temperature are quoted for reference; they are not components of the Beer Lambert Law equation).

Question 1: The absorbance of a solution of NADH measured at 340 nm is 1.12.

Calculate the concentration of the NADH solution.

Using the Beer Lambert Law: Absorbance (A) $= L \in C$

$$1.12 = 1.0 \times 6220 \times C$$

$$C = \frac{1.12}{6220}$$

$$C = 0.00018 \, M$$

Answer: 0.00018 M or 0.18 mM or 180 µM (see Chapter 1 for information on the concentration of solutions).

Question 2: The absorbance of a solution of NADH measured at 340 nm is 0.34. Calculate the concentration of the NADH solution.

Using the Beer Lambert Law: Absorbance (A) $= L \in C$

$$0.34 = 1.0 \times 6220 \times C$$

$$C = \frac{0.34}{6220}$$

$$C = 0.000054 \, M$$

Answer: 0.000054 M or 0.054 mM or 54 µM (see Chapter 1)

Figure 3.7 A simple spectrophotometer.

3.7 Spectrophotometers

3.7.1 Introduction

A variety of instruments have been manufactured to measure the absorbance of light by liquids.

Colourimeters (instruments that measure coloured reagents) are relatively unsophisticated machines that use:

- prisms to refract white light into its constituent colours,
- filters to abstract interfering wavelengths of light,
- a cuvette to hold the sample and galvanometers to record the intensity of the transmitted light.

Spectrophotometers (see Figure 3.7) are instruments which usually consist of:

- a light source (sometimes two),
- a monochromator (a device used to generate light at different wavelengths),
- a sample chamber with lid, containing a cuvette to hold the sample (possibly temperature controlled),
- a photomultiplier to detect the transmitted light
- a data analysis software package.

In most spectrophotometers the light absorbed by the solution the sample is dissolved in (the background or reference) is subtracted from absorbance of the sample. In single beam instruments this is achieved by measuring the absorbance of reference solution in a cuvette (sometimes called 'the blank') and setting the absorbance to zero before measuring the absorbance of the sample. The zero must be reset if the wavelength is changed. Dual beam spectrophotometers have two cuvette holders; one for the reference solution and one for the sample. Light at the required wavelength is directed intermittently on the reference and sample cuvette by means of a rotating mirror. In this configuration the absorbance of the sample can be constantly compared with the reference solution which is helpful if the wavelength is changed. A pair of optically matched cuvettes may be needed for accurate measurements.

Diode array spectrophotometers have a different optical arrangement to conventional spectrophotometers. The sample chamber is open as light is monochromated by a fixed angle diffraction grating after leaving the sample. The entire spectrum of monochromated light is detected by an array of photosensitive diodes which means no wavelength changing mechanism is required. However, wavelength resolution is limited to 2 nm increments by the use of these diodes.

3.7.2 The Light Source

To cover the complete range of the UV/light area of the electromagnetic spectrum (200–800 nm), two different light sources are required. A deuterium arc lamp provides UV light from 200–360 nm and a tungsten filament lamp, which provides light from 360–800 nm. The outputs of these bulbs complement each other as there is little overlap from either bulb in the other's wavelength range. In general, the deuterium lamps are more expensive to replace and have a shorter life span than the tungsten bulbs. Remembering to switch the deuterium bulbs off after the instrument has been used can lower the running costs. Xeon flash lamps generate a wide range of wavelengths covering both the UV and visible ranges and they can reduce the cost because they are only used when the sample is being measured.

3.7.3 The Optics Used to Generate Wavelengths of Light

White light from the light source is first collimated (see Figure 3.7) into a parallel beam by a curved mirror before being refracted by a prism into its constituent colours. However, a drawback with prisms is that the shorter wavelengths of blue light are dispersed more than the longer wavelengths of red light due to the density of the prism. A diffraction grating[4] consists of a silica disc with many finely scored lines ($2000 + mm^{-1}$) etched onto its surface. The grating diffracts light into many parallel beams travelling in different directions. The direction of the beams depends on the wavelength of light and spacing of the lines on the diffraction grating. The advantage of diffraction gratings is that they can resolve all the wavelengths of light (from blue to red) independent of the wavelength number. The diffraction grating can then be angled to ensure that the selected wavelength of light impinges on the sample to be measured.

3.7.4 The Slit Width

The light leaving the diffraction grating exits through a slit before encountering the sample in the cuvette and ultimately the detector. If the width of the slit is increased beyond 1.0 nm this will result in a decrease in the observed absorbance. Most spectrophotometers with diffraction gratings will have a fixed exit slit width to provide a controlled spectral band width.

3.7.5 The Cuvette

The device to hold the sample in the light beam of the colourimeter or spectrophotometer is called a cell or a cuvette (see Figure 3.8). When light is measured in the visible wavelengths of the electromagnetic spectrum, the cuvette can be made from glass, quartz or a non absorbing plastic. But because glass absorbs light in the UV wavelengths of the electromagnetic spectrum, cuvettes

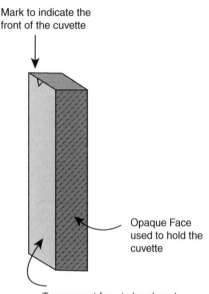

Mark to indicate the
front of the cuvette

Opaque Face
used to hold the
cuvette

Transparent face to be placed
in the light beam of the spectrophotometer

Figure 3.8 A diagram of a spectrophotometer cell/cuvette.

made of quartz or special plastics must be used. Optically matched pairs of cuvettes should be used in dual beam instruments.

The cuvette is the device that provides the interface between the sample and the spectrophotometer. This means that, if it is used incorrectly, it could be a major source of inaccurate measurements.

Cuvettes have a clear face and an opaque face (see Figure 3.8). They may also have a mark to indicate the front and the back of the clear face. The cuvette is placed into a sample holder to ensure reproducible alignment with the light beam. Then the lid should be closed to prevent stray light from interfering with the readings.

- In teaching laboratories, the two most common volumes for cuvettes are 1.0 ml and 3.0 ml with a 1.0 cm path length. Other cuvettes are available for more specialized applications.
- Always hold the cuvette by the opaque face and place the clear face in the path of the spectrophotometer light beam with the mark facing the light beam.
- Older cuvettes may have imperfections (small scratches). This is not usually a problem if the same cuvette is used for both the reference solution (blank) and

the sample solution. In addition, the cuvette should always be presented to the light beam in the spectrophotometer in the same orientation for each measurement.

- Check that the transparent surfaces are clean. If required wipe the surface with a tissue before use.

- Most spectrophotometers will accurately measure the absorbance between 0 and 2.0 (sometimes 0–3.0; see Section 3.5). Concentrated samples with an absorbance outside this range will register *off scale*. When you place the sample liquid in the cuvette, check to see if you can see through the sample. If you cannot see through the sample, the spectrophotometer will also struggle to read the absorbance.

- Check to see if there are any floating particles in the sample. These floating particles may deflect the light preventing the light from reaching the detector (attenuance) giving an inappropriate reading. In addition, the particles may move through the light beam on convection currents generated in the cuvette by the heat of the machine. In this case the readings will rise and fall as the particles move through the light beam. These will attenuate the light beam and be a source of error. If there are particles in the sample, remove the liquid and filter or centrifuge (see Chapter 5) to remove the particles in the sample before placing it again in the spectrophotometer.

- Atmospheric gases dissolve more in cold liquids and will be released as the temperature rises. If cold reagents are placed into the spectrophotometer the released gases may form bubbles which may pass through the light beam causing fluctuations in the reading. If this is suspected of happening take the cuvette out and gently tap the cuvette on the bench to release the bubbles from the inner surface of the cuvette. Allow the reagents to warm up to operating temperature and then replace the cuvette in the spectrophotometer.

- In addition, cold reagents placed in a warm machine may result in condensation on the cuvette's surface generating anomalous results. If this happens remove the cuvette, dry the surface with a tissue and replace in the machine.

- More sophisticated machines will have temperature controlled cuvette holders.

KEY POINTS TO REMEMBER 3.5

Glass absorbs light in the UV region. Quartz must be used for prisms, gratings and cuvettes (disposable plastic cuvettes are also available) when measuring the absorbance of a liquid in the UV region of the EM spectrum.

3.7.6 The Detector

The light that has not been absorbed by the sample solution passes through the cuvette and travels to the detector. Photomultipliers are routinely used in spectrophotometers as detectors. The photons of light strike a photocathode which releases electrons in proportion to the photons striking the surface. The released electrodes strike dynodes set up in a series of increasing potential. When the electrons from the photocathode strike the first dynode it causes a secondary release of electrons, which is enhanced with each subsequent dynode. The electron density reaching the anode from the final dynode will be multiplied but remains proportional to the original electron emission from the photocathode. The light intensity is reflected by the amount of current generated and this signal is processed for viewing on the readout of the spectrophotometer.

3.8 Applications of Spectrophotometry in Bioscience

3.8.1 Direct Measurements of Biological Molecules

As mentioned in the introduction, spectrophotometers can be used to measure the concentration of many common biological molecules. Some of the macro-molecular structures such as proteins and nucleic acids can be measured directly. Proteins characteristically absorb light at 280 nm due to the presence of the aromatic amino acids (tryptophan and tyrosine) in their structure. Nucleic acids characteristically absorb light at 260 nm due to the presence of nucleotides (cytosine, adenine, thymine, uracil and guanine) within their structure.

For a quick method to measure the concentration of proteins in solution place the solution in a spectrophotometer and measure the absorbance at 280 nm. For a pure protein solution a 1% (w/v) absorptivity ($\varepsilon^{1\%}$) value at 280 nm can be determined and the acquired reading gives a direct measure of the protein's concentration. This method has problems when used with samples derived from homogenized samples (see Chapter 5). The crude homogenates will also contain metabolites, nucleic acids and amino acids which will also absorb light at 280 nm. The interference may be reduced by dialysis to remove small molecular weight metabolites before reading the sample in the spectrophotometer. In addition, a correction factor can be used if nucleic acids are considered to be still present.

$$\text{Protein}(\text{mg ml}^{-1}) = 1.55(A_{280\ nm}) - 0.76(A_{260\ nm})$$

Alternatively, a standard calibration graph (see Chapter 4) of known concentra-tions of bovine serum albumin (X-axis) can be plotted against the absorbance at 280 nm (Y-axis). The absorbance values of samples can be read from the standard graph and converted into a protein concentration value. The problem with this method is that not all proteins contain the same number of tyrosine and tryptophan residues as bovine serum albumin and the concentration of some proteins will be over or under estimated.

The purity of a sample of DNA can also be determined by measuring the absorbance at both 260 nm and 280 nm. A ratio between 1.7 and 2.0 represents good quality DNA. If the sample is turbid an additional correction factor can be included by also measuring the absorbance at 320 nm.

$$\text{DNA purity}(260\,\text{nm}/280\,\text{nm}) = \frac{(\text{A260 nm} - \text{A320 nm})}{(\text{A280 nm} - \text{A320 nm})}$$

Good quality single stranded RNA produces a 260 nm/280 nm ratio of 1.9–2.1 in 10 mM Tris buffer pH 7.5 (the pH influences this ratio).

3.8.2 Chromophore Assays

Very few metabolites in biology absorb light in the visible range of the electromagnetic spectrum. This could be problematic if a spectrophotometer with only a tungsten light source for measuring visible light is available. Over the years, methods have been developed to attach reagents to target molecules to produce a coloured product which can then be measured in the visible light range of the electromagnetic spectrum.[5] Ideally the reagent should react with a limited group of compounds to provide selectivity and form the basis of a quantitative assay (see Table 3.1). If the molar absorptivity coefficient ($\in$) of the coloured product is known, this can be used to determine the concentration of the metabolite in the sample. Alternatively, a standard calibration graph can be prepared (see Worked Example 3.3 and Chapter 4).

Using this approach there can be drawbacks measuring samples derived from crude cell homogenates. Metabolites may be present which will also react or partially react with the reagent making it difficult to determine accurate results.

Table 3.1 *Some popular colourimetric assays*

Detection reagent	Target analyte	Detection wavelength (nm)
Ninhydrin	Amino acids	570
Coomassie blue	protein	595
Diphenylamine	DNA	595
Orcinol (Bial's reagent)	RNA	665
Ammonium molybdate	Inorganic phosphate	600
Folin's phenol reagent	Tyrosine, Phenols	660
3,5-dinitrosalicylic acid	Reducing sugars	480, 570

Appropriate sample preparation may be required to remove contaminating molecules to increase the accuracy of the results.

Worked Example 3.3: To determine the free amino acid concentration in a crude cell homogenate

The tissue should be homogenized and clarified (see Chapter 5). The addition of ice cold trichloroacetic acid to a final concentration of 7% (w/v) will precipitate any proteins in the supernatant. This precipitate can be removed by centrifugation at 13 000 g for 30 min at 4 °C. The protein free supernatant will contain small molecular weight metabolites including the free amino acids.

- Prepare a solution of alanine for the standard calibration graph in 7% (w/v) trichloroacetic acid.
- Prepare a reference (blank). For this example it would be prudent to use 1.0 ml of 7% (w/v) trichloroacetic acid.
- Prepare 1.0 ml dilutions of the standard alanine in triplicate
- Prepare 1.0 ml dilutions of the sample in triplicate
- Add 0.5 ml 0.1% (w/v) ninhydrin and boil for 5 minutes.
- An intense blue colour will develop.
- Cool the standards and samples to room temperature and set wavelength number on the spectrophotometer to 570 nm.
- Use the reference sample to set the absorbance to zero.
- Measure the absorbance of the standards and samples.
- Prepare a calibration graph by plotting the known concentration of standard alanine (x-axis) against the recorded absorbance for each standard (y-axis).
- Use the calibration graph to convert the absorbance of the diluted sample into an amino acid concentration. One of the dilutions should fall on the linear part of the calibration graph and remember to multiply up by the dilution factor (see Chapter 4).

Note: At high concentrations the graph may deviate from linearity and form a plateau. This may be because the reagent is not present at a high enough concentration to produce sufficient colour or it may be due to the formation of interactions between molecules. Either way, the Beer Lambert Law is only valid in the linear part of the graph.

If the absorbance of the sample does not fall on the linear part of the graph, the sample should be diluted and the reaction repeated until the absorbance reading does fall within the linear part of the graph.

3.8.3 Enzyme Catalyzed Reaction

Spectrophotometry can be used to follow the course of an enzyme catalyzed reaction. For example, dehydrogenase enzymes use the cofactors NAD^+/NADH. Thus, the appearance or disappearance of NADH can be quantified by following the course of the reaction at 340 nm (see Figure 3.4b).

Worked Example 3.4: Lactate dehydrogenase assay

pyruvate   lactate

The enzyme lactate dehydrogenase catalyzes the conversion pyruvate to lactate; as part of the reaction the cofactor NADH is oxidized to NAD^+. The progress of the reaction (the rate of disappearance of NADH) can be monitored by recording the change in absorbance at 340 nm (ΔA_{340nm}) in a spectrophotometer.

Into a 1.0 ml cuvette was placed 0.9 ml of 200 mM Tris/HCl buffer pH 7.4, 0.05 ml of 6.6 mM NADH and 0.05 ml of 30 mM sodium pyruvate. The cuvette was placed into a spectrophotometer set to record the absorbance at 340 nm. The non enzymic background change in absorbance was measured over a 5 minute period.

The cuvette was emptied, rinsed and refilled with 0.8 ml of 200 mM Tris/HCl buffer pH 7.4, 0.05 ml of 6.6 mM NADH, 0.05 ml of 30 mM sodium pyruvate and the reaction started by the addition of 0.1 ml homogenized and clarified (see Chapter 5) bovine heart muscle (diluted to 0.5 mg ml^{-1} prior to use in 200 mM Tris/HCl buffer pH 7.4).

The following data were obtained:

Time (min)	0.0	1.0	2.0	3.0	4.0	5.0
Absorbance 340 nm	0.8	0.68	0.56	0.48	0.41	0.35

- Remember that you are measuring a rate so it does not matter too much where the initial absorbance reading starts from.
- If the absorbance reading changes very quickly over a 1.0 min period repeat the experiment with a more dilute enzyme preparation.

- Some spectrophotometers will determine the rate and present the data in a table. Otherwise a graph of A_{340nm} against time can be plotted to determine the gradient of the graph that is $\Delta A_{340nm}\ min^{-1}$ (remember to use the initial *linear* part of the graph over a sensible time period).

The molar absorptivity coefficient ($\in$) of NADH at 340 nm is 6.22×10^3 $(M^{-1}\ cm^{-1})$ at 25 °C and the path length of the cuvette is 1.0 cm.

The change in absorbance at 340 nm over a 1 minute period between 1.0 and 2.0 minutes is 0.12.

Using the Beer Lambert Law: Absorbance $(A) = L \in C$

$$0.12 = 1.0 \times 6220 \times C$$

$$C = \frac{0.12}{6220}$$

$$C = 0.000019\ M\ min^{-1}$$

$$= 19\ \mu M\ min^{-1}$$

$$= 19\ \mu mol\ in\ 1000\ ml\ min^{-1}$$

The Beer Lambert Law provides the concentration in mol litre^{-1} (see Chapter 1).

The amount of product in the 1.0 ml cuvette can be found by dividing by 1000

$$= \frac{19}{1000}\ \mu mol\ min^{-1}$$

$$= 0.019\ \mu mol\ min^{-1}$$

$$= 19\ nmol\ of\ product\ formed\ min^{-1}\ in\ the\ 1.0\ ml\ cuvette$$

International units of enzyme activity:

1 unit = 1 μmol of product formed min^{-1}.

There are 0.019 units of lactate dehydrogenase in the 0.1 ml of the bovine heart extract or 0.19 units in 1.0 ml of bovine heart extract (0.5 mg ml^{-1}).

$$Specific\ activity(units\ mg^{-1}\ protein)\ =\ \frac{units\ ml^{-1}}{protein\ (mg\ ml^{-1})}$$

$$=\ \frac{0.19}{0.5}$$

In this experiment there are 0.38 units of lactate dehydrogenase mg^{-1} of bovine heart protein.

3.9   Summary

- Light in the UV/visible light (200–800 nm) part of the electromagnetic spectrum can be measured in a spectrophotometer.
- The absorbance of light is related to the concentration of a solution and the path length the light has to travel through the solution.
- The biological molecules in solution can be detected and quantified using a spectrophotometer and the Beer Lambert Law.

Notes

[1] For example, compounds that fluoresce have a wavelength number at which they maximally absorb photons (excitation wavelength) and a wavelength at which they re-emit a photon (emission wavelength). This is described in Section 3.3.1.

[2] Worked Example 3.1 demonstrates that longer wavelengths of the electromagnetic spectrum have less energy.

[3] Molar concentrations are not always suitable for proteins and 1% (w/v) absorptivity at a wavelength at 25 °C is quoted (for example, bovine serum albumin (BSA) at 280 nm $\varepsilon^{1\%} = 6.6$).

[4] The surface of a CD or DVD is etched with many closely spaced reflective lines. These discs can act as a diffraction grating; if they are angled in a beam of white light the colours of the rainbow can be seen.

[5] A chromophore is the part of the molecule which is responsible for the colour.

4

DATA ANALYSIS AND PRESENTATION

4.1 Introduction

A good laboratory report is one that contains data that has been analyzed appropriately and presented effectively. This chapter gives you an introduction to some of the key elements of basic statistical analysis applied to data in the biosciences and provides hints and tips on how to present your data effectively in laboratory reports.

Although you may do some experiments in laboratory classes only once, it is important to realize that you would be unlikely to get exactly the same result if you repeated the experiment. This is because, with biological measurements, there is always variation or spread of data. For this reason, it is important to establish the reliability of data in terms of the accuracy and precision of the values obtained from bioscience experiments.

Definitions: Accuracy and Precision

Accuracy
- This refers to the closeness of a measured value (or an estimate from a collection of measurements) to the actual value.
- In the laboratory this can be determined by comparing the experimental output from the test or assay in question with those of a reference method.

Precision
- This refers to the repeatability of the experimental measurements, as measured by the spread of data which reflects the reproducibility of measurements.
- Ideally, an experiment should produce readings that are both accurate and precise.

Worked Example 4.1: Activity Measurements

If the specific activity of an enzyme measured by a long established method is 12 units/mg protein, which would be the most suitable alternative based on

---

Basic Bioscience Laboratory Techniques: A Pocket Guide, First Edition.
Philip L.R. Bonner and Alan J. Hargreaves.
© 2011 John Wiley & Sons, Ltd. Published 2011 by John Wiley & Sons, Ltd.

the measurements from three potential new assay methods (A, B and C) giving the specific activities indicated below: –

a. 12.4 ± 0.6
b. 12.1 ± 0.1
c. 12.0 ± 0.6

- Method A is unsuitable as it gives values that are inaccurate and imprecise.
- Although the mean value from method C is the most accurate, it is not the most precise as indicated by the wide spread of data.
- Method B may be the most suitable alternative from the three methods because it gives values within 1% of the expected values (accurate) with a very narrow spread of data (precise).

Effective assessment of data requires statistical analysis, which could take various forms including basic descriptive statistics of data (e.g. estimation of midpoint and spread of data), to assessment of the relationships between different variables or sets of data. However, for effective communication of your findings, it is also important to present your data clearly and unambiguously in your laboratory report.

Properly used, statistical analysis provides valuable support for the scientific validity of findings. However, improperly applied statistical analysis can be misleading as indicated by the numerous quotes you will see at discrete points in this chapter.

Statistics

"There are three kinds of lies: lies, damn lies and statistics" - Attributed to Benjamin Disraeli by Mark Twain.

"Facts are stubborn but statistics are more pliable" - Mark Twain.

"Definition of statistics: The science of producing unreliable facts from reliable figures" – Evan Esar.

- These quotes refer to the common skepticism at the use of statistics to justify political decisions or to criticize those of opponents.
- This reflects the selective use of statistics to support a specific argument rather than the appropriate use of statistics to analyze the validity of data.
- The moral of the story is that an appropriate statistical analysis should be applied to the data in question and the method used should be declared openly and honestly in any publication or presentation using such data so that it can be open to scrutiny.

- In most areas of science there is not always agreement on the methods used to analyze data.

"Definition of a statistician: A man who believes figures don't lie but admits that under analysis some of them won't stand up either" – Evan Esar.

4.2 Statistical Analysis of Data: Some Key Definitions

4.2.1 Populations and Samples

In statistics, the term "population" refers to very large sets of data that we would ideally want to analyze (i.e. large numbers of measurements). However, in reality it is simply not practical to analyze large populations. Instead, a *sample* is studied, which contains a smaller number of data points "picked" from the population, which is assumed to be representative (i.e. an estimate) of the whole population.

4.2.2 Variables and Observations

A *variable* is a characteristic that differs from one member of the population and the next and they can take various forms such as qualitative (e.g. observed or not observed) or quantitative (e.g. specific activities of an enzyme, protein concentration etc.).

Different types of statistical analysis are required to analyze samples containing these distinct types of data. As a minimum requirement, some form of descriptive statistics should be performed on your data.

4.2.3 Descriptive Statistics

This term refers to a number of analyses that can be applied to describe the midpoint and spread of data within a sample (see Worked Example 4.2). Estimates of midpoint include:

Median - This is the value in the middle when numbers have been listed in rank order.

Mode - This is the most frequently occurring observation.

Mean - The mean is the arithmetic average of all values in the sample or population. To calculate it we have to add all values together and divide the sum by the number of values included.

Worked Example 4.2: Estimation of Mid Point Values

Median
- Calculate the median of the set of values 1, 2, 3, 4, 5.
- To do this, look for the obvious mid point in the values, if one exists. In this case if the values are placed in increasing or decreasing order the clear mid point value is 3.
- If the values were 2, 4, 8, and 10, the median would be 6 (in this case, the two middle numbers in the group are averaged).
- Thus, in the event of an even number of values in the sample, the median would be the average of the middle two values.

Mode
- Estimate the mode in the following set of values: 1, 5, 3, 5, 7, 5, 5, and 8.
- To do this simply measure the frequency of each value and the one that appears most frequently is the mode.
- In this case the value 5 appears four times and is clearly the mode, as the other values only appear once.

Mean
- Estimate the sample mean for the following data: 23, 13, 20, 32, 5, 10, 7, 25, 11, and 18.
- To do this, calculate the sum of the ten values, then divide by the number of values (10).
- $= 164/10 = 16.4$.
- As discussed earlier, this value would effectively be our estimate of the population mean, which would apply to a much larger number of data values.

4.2.4 Measures of Spread (Dispersion)

Different sample populations may have the same mean, median or mode but differ in the *spread* of data. Variability is likely to be greater in small sample sizes and, as the mean of the sample is an estimate of the population mean, an estimate of the variability of the sample data can give us an idea of the accuracy and precision of the calculated mean value.

There are various measurements of variation of relevance to bioscience research data, which are indicated below. These include:

Range: Highest and lowest values, and so on.
Variance: (S^2) can be determined from the following equation:

$$S^2 = \frac{1}{n-1} \sum_{i=1}^{n} (x - \bar{x})^2$$

Where n represents the number of measurements, i is the ith value, x is each individual value and $\bar{x}$ is the mean of all measurements. This is an estimate of the *population* variance, which is normally referred to as σ^2.

Standard deviation (s): This is the square root of the variance ($\sqrt{s^2}$).
One formula to estimate the sample standard deviation (s) is:

$$s = \sqrt{\frac{1}{n-1} \sum_{i=1}^{n} (x - \bar{x})^2}$$

The sample standard deviation is an *estimate* of the population standard deviation ($\sqrt{\sigma^2}$). For each data point, deviation from the mean is calculated as the difference between that value and the mean, which could be a positive or negative value. Typically, standard deviation values are represented as error bars for (mean) values of data presented in graphs, histograms. Alternatively data may be presented in tables as mean $\pm$ SD.

Alternatively, *for small sample sizes*, the standard error of the mean (SEM) is often used to estimate the spread of data. This is derived by dividing the standard deviation by the square root of the numbers of values, as shown below:

$$SEM = \frac{s}{\sqrt{n}}$$

Such values give an indication of the spread of a population's data about its mean. As a general rule the spread of data decreases as the population size increases. Thus, small sample sizes may result in a broader distribution curve, reflecting wider standard deviations.

You should try to calculate these values using the formulae above so that you can understand better the principles involved (see Worked Example 4.3). However, there is a variety of statistical packages for PCs that can be used to perform basic descriptive statistics (e.g. Microsoft Excel, Graphpad Prism, Minitab, etc.).

Worked Example 4.3: Calculation of Mean and Standard Deviations

Find the mean and standard deviation from the following set of values (x): 23,13,20,32,5,10,7,25,11,18.

To do this, construct a table with the headings x, $x-\bar{x}$ and $(x-\bar{x})^2$ and fill in the individual values given above, their differences from the mean (which may be positive or negative values) and the square of the mean differences (always positive).

- These values can then be substituted into the equations above to calculate the standard deviation.

X	$x-\bar{x}$	$(x-\bar{x})^2$
23	6.6	42.56
13	−3.4	11.56
20	3.6	12.96
32	15.4	237.16
5	−11.4	129.96
10	−6.4	40.96
7	−9.4	88.36
25	8.6	73.96
11	−5.4	29.16
18	1.6	2.56
	TOTAL	669.2

- Calculate the mean by summing all of the values and dividing by the number of values.
- Mean $(\bar{x}) = \frac{(23+13+20+32+5+10+7+25+11+18)}{10} = \mathbf{16.4}$
- Now calculate the standard deviation, incorporating the calculated value from the sum of the squares of the differences from the mean (i.e. 669.2)

$$s = \sqrt{\frac{1}{(10-1)} \times 669.2} = \sqrt{74.36} = 8.623$$

Standard deviation = **8.623**
- The standard error is then obtained by substituting the standard deviation value in the following equation:
- SEM = SD/$\sqrt{n}$ = 8.623/3.162 = **2.727**

4.3 Distributions

While basic descriptive statistics will give us an idea of the midpoint and spread of data points within a sample, in order to establish the significance of any differences observed between basic data from two or more different samples, it is necessary to apply an appropriate statistical test. The type of test that should be applied is, in turn, dependent on the nature of the distribution of values within the samples. Several types of distribution exist as discussed below.

4.3.1 Normal Distribution

In a population with a normal distribution, the individual values are distributed symmetrically about the mean, giving a bell-shaped curve (Figure 4.1a). In theory, a perfect normal distribution is only likely to occur in large samples and populations, although it is possible to determine the likelihood that a smaller sample approximates to a normal distribution. In practice the normal distribution is converted to a standard normal distribution (or z *distribution*) by introducing a

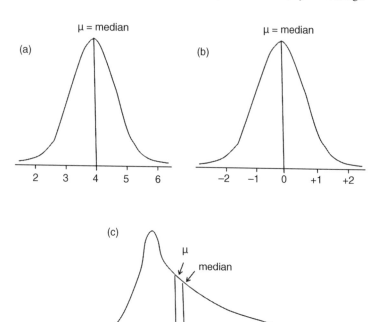

Figure 4.1 *Examples of normal and non-normal distributions. Shown are (a) a normal distribution, (b) a standard normal (or z) distribution and (c) a non normal distribution.*

correction factor to give each sample population a mean of zero, using the formula:

$$z = \frac{X - \mu}{\sigma}$$

Where X is an individual score, μ is the mean and σ is the standard deviation of the original normal distribution. In the z distribution, 96% of values lie within the mean ± 2 SD units and 68% within the mean ± 1 SD unit (Figure 4.1b). Therefore, calculation of the z value (see Worked Example 4.4) for a particular measurement will give an indication of by how many standard deviation units it differs from the population mean. If the value is more than 2 SD units above or below the mean, it falls outside the 96% range.

Worked Example 4.4:

Find the Z value for the mark of a student scoring 35 in a test for which the average mark was 25 and the standard deviation was 5.

$$Z = (35-25)/5 = 2$$

- This means that the mark obtained was the equivalent of two standard deviation units higher than the class average.
- This value together with z values for the other students' marks would be used to construct the standard normal distribution of the class marks.

A randomly drawn element from the population has a 4% probability of falling within the outer range. As it is typical to apply confidence limits of 95%, the 95% confidence interval can also be expressed, which represents the range within which 95% of values will be located if the mean of the sample is not significantly different from the population mean (μ). In general, the narrower the calculated range, the higher the level of confidence in the accuracy of the estimation.

The confidence interval (CI) is calculated as follows:

CI = mean *plus or minus* t value x standard error

Confidence interval calculations can form an important part of more complex statistical tests such as analysis of variance (ANOVA), which can be used more effectively for the comparison of multiple sets of data but will not be discussed here. However, an example of confidence interval estimation to determine whether sample means are likely to be different from each other is shown below. If the confidence intervals do not overlap the means are likely to be significantly different from each other.

This example contains data sets from an experiment in which the effects of a toxin were studied on the outgrowth of axons by cultured nerve cells at two distinct time points (4 and 24 h).

It should be noted that, in the worked example, the wider spread of values in toxin treated cells means that a more robust statistical test should be applied, as discussed later.

Worked Example 4.5: Confidence Intervals

This example contains data sets from an experiment in which the effects of a toxin were studied on the outgrowth of axons by cultured nerve cells at two distinct time points (4 h and 24 h).

Determine the confidence intervals for the data below.

- Statistical packages may provide a plot of these: If they don't overlap, the means are likely to be different.
- However, if you check the probability level, they may be separated at 0.05, but not at 0.01, giving an idea of how significantly different the datasets are from one another.

Confidence intervals are the mean $\pm$ t x standard error, where the value of t is looked up in a statistical table for n-1 degrees of freedom and the desired level of probability. For example in the case of three replicates, n-1 = 2; for a probability of 0.05 (which sets confidence intervals to 95%), t = 4.3 (see Section 4.4.1).

For each column of data, the standard error of the mean or SEM = $\sqrt{s^2/n}$

Statistic	Control 4 h	Toxin-Treated 4 h	Control 24 h	Toxin-Treated 24 h
variance	19.59	7.47	16.00	5.13
SEM	2.21	1.37	2.31	1.13
t x SE	9.50	5.89	9.89	4.86
mean	32.35	13.23	35.9	18.5
Confidence interval at 95% limits	22.85 → 41.85	7.34 → 19.12	26.01 → 45.79	13.64 → 23.36

- Initial observation of these confidence limits in the bottom row of the table indicates that there is no overlap between treatments and control at a given time point.
- Therefore it is likely that the toxin is having an inhibitory effect on axon growth.
- These limits can be plotted manually or on statistical packages (see below for a typical plot in Minitab), to have a more visual confirmation of the lack of overlap.
- However the overlap between controls at 4 h and 24 h suggests that there is no significant increase in axon numbers in controls from 4 to 24 h incubation.

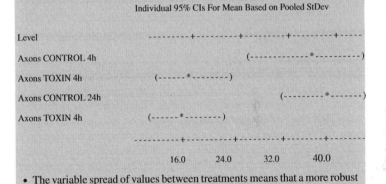

Individual 95% CIs For Mean Based on Pooled StDev

Level	
Axons CONTROL 4h	(-------------*----------)
Axons TOXIN 4h	(------*--------)
Axons CONTROL 24h	(---------*-------)
Axons TOXIN 4h	(------*--------)

16.0   24.0   32.0   40.0

- The variable spread of values between treatments means that a more robust statistical test should be applied, as discussed later.

4.3.2 Asymmetric (Non-Normal) Distribution

In this type of distribution, the sample values are distributed in a more skewed fashion (Figure 4.1c), which is more typical in small sample sizes. For statistical analysis, samples showing such distributions should be compared with each other about their mid point, in this case the median. Statistical analysis to determine the similarity between two samples with non-normal distributions would normally involve the use of a non-parametric statistical test such the Mann Whitney U test or Kruskall Wallis test for multiple comparisons. These tests are not discussed in this chapter but you can find out more about their use in bioscience data analysis by looking at the suggested further reading at the end of this book.

4.3.3 Discrete Probability Distributions

The binomial distribution is another common type of distribution that you may come across in your early studies at university. This applies to a situation where there are two possible equally likely outcomes that are completely independent of each other (e.g. success/failure, survival/death, male/female, etc.) and have an equally likely chance of occurring if trials (i.e. the process of selecting an individual at random from the population) are done randomly, independently of each other and in sufficient numbers.

The Poisson distribution is important in the analysis of certain types of data associated with randomness. It is used in situations where nothing can influence the data except chance. In this case, the Poisson distribution would describe the number of counts we would expect per unit time or space and the data would present a Poisson distribution around the mean of the fitted values. The Poisson distribution is also discrete in that it is suitable for describing count data.

For example, the likelihood of obtaining a head after tossing a coin is, in theory, 0.5 (50%). In practice this means that if the coin is tossed a large number of times, the outcome should be 50% heads and 50% tails ($p = 0.5$). Similarly, in the case of rolling a dice, the probability of obtaining a specific value (e.g. 1) is 1 in 6, whereas the probability of obtaining any number but one on a given trial would be 5/6.

Probability

"If a coin falls heads repeatedly one hundred times, then the ignorant would claim that the 'law of averages' must almost compel it to fall tails next time, any statistician would point out the independence of each trail and the uncertainty of each outcome. But any fool can see that the coin must be double headed." – Ludwick Drazwk.

"The statistics on sanity are that one out of every four Americans is suffering from some kind of mental illness. Think of your three best friends. If they're OK, then it's you." – Rita Mae Brown.

"The buffalo isn't as dangerous as everyone makes him out to be. Statistics prove that in the United States more Americans are killed in automobile accidents than are killed by buffalo." – Art Buchwald.

"Statistics show that of those who contract the habit of eating, very few survive" – William Wallace Irwin.

So, as you can see, you shouldn't believe everything you read about statistics!

4.4   Statistical Comparison of Data

In order to determine whether sets of data are statistically different from each other, it is convention to apply the null hypothesis (H_0), which usually states that there is no difference between the data samples (i.e. they are likely to have arisen from the same population distribution - H_0: $\mu = 0$). The alternative hypothesis is that there is a difference, although the direction of that difference is rarely stated (i.e. H_0: $\mu \neq 0$). In principle, whatever test is applied (which depends on the precise nature of the data and its distribution), the null hypothesis is not rejected unless there is convincing evidence against it. Both the null and the alternative hypotheses are expressed in a manner that reflects hypothetical population parameters (e.g. as it is unlikely that the sample mean will be equal to zero). One tailed and two tailed tests can also be applied, depending on whether any differences from the mean are expected to occur in a specific direction (i.e. an increase *or* decrease) or whether they could occur equally in either direction, respectively.

It is typical to use 95% confidence limits when performing a statistical test in bioscience research. In other words, the calculated probability that the two samples belong to the same population must fall below 1 in 20 or 5% (normally expressed as $p < 0.05$) in order that the null hypothesis can be rejected. The lower the probability value falls below the 0.05 level, the more highly significant the difference. For example, a value of $p < 0.01$ would indicate that there is less than a 1% chance that the samples being compared could have arisen from the same population, and so on. On the other hand, a value above 0.05 would be taken as evidence that the two sample distributions could have resulted from chance variation within the population are not therefore likely to be significantly different from each other. However, if sample measurements in biomedical studies show significant levels of variability and the p value is approaching significance (for example 0.05 to 0.1), significance may be eventually achieved by increasing the number of measurements in each sample (particularly if the sample sizes are on the small side). Let's now look at the basis of some of the more common statistical tests used on bioscience data, and how they are applied to different sample distributions and data types.

4.4.1   Comparing Two Sets of Values

4.4.1.1   *Unpaired and Paired Parametric Tests (t Tests)*

This type of test (see Worked Example 4.6) is applicable to samples showing a normal distribution and is often applied in biomedical research studies. The unpaired t test is typically used to compare two independent samples about their arithmetic means. Prior to using this test, it is advisable to check that both sets of data are normally distributed, as this is not always the case for small sample sizes. If the test is applied to samples with non-normal distributions the confidence limits at which rejection of the null hypothesis is set may not be reliable. Thus, conclusions drawn from data showing significant or non significant changes with p values close to the chosen confidence limits may be wrong if the wrong significance test is applied. Having said that, the test may still work well when distributions deviate from normality, provided that they are reasonably symmetrical and continuous. Therefore, very low p values (e.g. 0.001) are still likely to reliably reflect that there exists a significant difference between two data sets.

The test is based on the t distribution of data. The t value for each data point can be obtained from the equation:

$$t = \frac{\bar{x} - \mu}{s / \sqrt{n}}$$

This is an approximation of the z distribution described earlier, taking into account the sample size. The shape of the t-distribution approximates to the z distribution for large sample sizes, whereas the curve becomes flatter at lower sample number (n). The calculated t value can be read off in statistical tables of t-distributions, which take into account the sample size and distribution by correcting for the 'degrees of freedom' for a particular set of data, which is determined by n-1 (where n is the number of measurements). This factor is normally taken into account when using a computer package to perform statistical tests. Otherwise you can look up the value in statistical tables, which can be found on line or purchased (see further reading section). With this test, you are effectively estimating the probability that the two samples could have arisen by chance from the same population distribution. A value of $p < 0.05$ would be indicative of a significant difference between the two sets of data.

Worked Example 4.6: Student's t Test

- Determine whether the following randomly taken set of measurements (12, 24, 13, 5, 10, 18, 8, 45, 14 and 13) could be part of a normal distribution with a mean of 6.
- First of all, estimate the *sample mean* as follows:

$$(12 + 24 + 13 + 5 + 10 + 18 + 8 + 45 + 14 + 13)/10 = 162/10 = \textbf{16.2}$$

If we then estimate the *standard deviation*, as described in the earlier worked example:

86

X	$x-\bar{x}$	$(x-\bar{x})^2$
12	−4.2	17.64
24	7.8	51.84
13	−3.2	10.24
5	−11.2	125.44
10	−6.2	38.44
18	1.8	3.24
8	−8.2	67.24
45	28.8	829.44
14	−2.2	4.84
13	−3.2	10.24
	TOTAL	**1158.6**

Square of mean differences

- The standard deviation of the sample is calculated as follows:

$$s = \sqrt{\frac{1}{(10-1)} \times 1158.6} = \sqrt{128.73} = 11.35$$

The t value is then calculated as follows:

$$t = \frac{\bar{x}-\mu}{s/\sqrt{n}}$$

$$= (16.2-6) \div (11.35 \div \sqrt{10})$$

$$= 10.2 \div 3.639$$

$$= 2.802$$

- If we look at a table of t distributions we can see that the 5% point for t is 2.26 at 9 degrees of freedom (n-1), which means that 95% of the distribution will lie with a t value of −2.26 to +2.26.
- As the observed t value falls outside the central 95% of the distribution, we can *reject* the null hypothesis at the 5% level, as we cannot reasonably assume that the sample was drawn from a population with a true mean of 6.0.

In order to compare two *sample* means, a similar approach is adopted taking into account the central limit theorem. Thus, if two independent biological samples (with means $\bar{x}$ and $\bar{y}$, containing N_1 and N_2 values respectively) are taken from a population with a normal distribution, they too are assumed to be normally distributed with a variance of $\frac{\sigma^2}{N_1}$ and $\frac{\sigma^2}{N_2}$.

The difference $(\bar{x}-\bar{y})$ contains two varying elements (i.e. the two sample means) both of which contribute to its total variance, such that the variance for $(\bar{x}-\bar{y})$ is equal to $\frac{\sigma^2}{N_1} + \frac{\sigma^2}{N_2}$.

We can then propose the null hypothesis that both $\bar{x}$ and $\bar{y}$ come from the same parent population (with mean and variance of σ^2 and μ), in other words that they are not significantly different from each other. In order to do this the pooled variance must first of all be calculated from the equation:

$$ S^2 = \frac{1}{N_1 + N_2 - 2} \times \left[\left(\sum x - \bar{x}\right)^2 + \left(\sum y - \bar{y}\right)^2 \right] $$

Then calculate the t value for $(N_1 + N_2 - 2)$ degrees of freedom from:

$$ t = \frac{\bar{x}-\bar{y}}{\sqrt{s^2 \left(\frac{1}{N_1} + \frac{1}{N_2}\right)}} $$

The t value as before can be read off (as explained earlier) in statistical tables for the appropriate number of degrees of freedom to determine whether the two sample means lie within the same distribution.

Alternatively, if the two sets of data to be compared are *dependent* on each other, a *paired t test* can be performed. In this case the null hypothesis is that the mean of the differences between the paired sets of observations is zero (i.e. that the distribution of mean differences is not significantly different from that of a population with a mean of zero). In biomedical research, an example of paired data could be the use of one set of patients to test the effects of two different drugs on a specific parameter such as blood pressure.

$$ t = \frac{\bar{d}-\mu d}{sd} $$

In this equation, $\bar{d}$ is the mean of the differences between paired observations, μd is the hypothesized mean difference between paired observations (usually zero), and sd is the standard error of $\bar{d}$. As for unpaired t tests, in the paired sample t test the value for degrees of freedom is n-1, with n representing the number of *pairs* of observations. A one sample t test is applied, as we have effectively reduced the data to one variable by dealing with the difference between each paired observation. For the test to be a valid indicator of significance, it is assumed that the distribution of the differences between individual pairs is normal and that the data were randomly selected from the distribution.

4.4.2 Unpaired and Paired Non-Parametric Tests

Most of your current needs for statistical testing will probably be covered by one or a combination of the tests described so far. However, the Mann Whitney U test is often applied to unpaired data sets that exhibit deviation from a normal distribution, which is often the case with small sample sizes. It compares the samples' distributions about the median rather than the arithmetic mean in order to take into account the asymmetric spread of the data. Although paired t tests are often used irrespective of the mean difference distribution, if the data sets are paired but exhibit a non-normal distribution, another non-parametric test should be applied, such as the Wilcoxon paired sample test. This test checks the number and sizes of positive and negative differences between paired data by putting them in rank order, summing the ranks of the positive and negative differences. The rank sums provide the test statistic which is then subjected to a binomial test to determine the p value. The null hypothesis is that there is no difference between the samples; therefore the sum of the positive and negative differences should be zero. This test can be applied using the One-sample Wilcoxon test, which is available in most statistical software packages.

4.4.3 Chi Square Test for Dichotomous Variables

Binomial distributions are distinct two-class populations (e.g. success/failure; dead/alive; sick/well; male/female). In this case, comparison can be made using the chi square test, which compares observed with expected outcomes. In a simple chi square test, the data is set out in tabular form:

GROUP	Success	Failure	Total
A (observed)	a	b	a + b
B (expected)	c	d	c + d
TOTAL	a + c	b + d	n

$$\text{Chi Squared } (\chi^2) = \frac{(ad-bc)^2 \times n}{(a+c)(b+d)(a+b)(c+d)}$$

In this particular case, the final chi squared value must be 3.84 or above to indicate a significant difference at $p < 0.05$ and 6.64 or above for $p < 0.01$. These values can be found in statistical tables and vary according to the number of degrees of freedom, which is in turn related to the number of measurements. As the table above contains two measurements (success or failure), the number of degrees of freedom $= 1$ (i.e. n-1). This is taken into account when reading chi square values in statistical tables, although a computer package would do this automatically.

Worked Example 4.7: Chi Square Test

A coin was tossed 100 times, with the following results (observed counts). Use the chi square test to determine whether these counts differ significantly from the expected counts (i.e. equal numbers of heads and tails). The null hypothesis would be that there is no significant difference between the two sets of counts (i.e. the coin is not biased towards heads or tails).

	Heads	Tails	Total
Observed counts (A)	54 (a)	46 (b)	100 (a + b)
Expected counts (B)	50 (c)	50 (d)	100 (c + d)
TOTAL	104 (a + c)	96 (b + d)	200 (n)

$$\chi^2 = \frac{(ad-bc)^2 \times n}{(a+c)(b+d)(a+b)(c+d)}$$

$$\chi^2 = \frac{(ad-bc)^2 \times n}{(a+c)(b+d)(a+b)(c+d)}$$

$$\chi^2 = \frac{(2700-2300)^2 \times 200}{104 \times 96 \times 100 \times 100}$$

$$\chi^2 = 32000000/99840000 = \mathbf{0.321}$$

Degrees of freedom = 1

In order to represent a significant difference from the expected value at $p \leq 0.05$ or $p \leq 0.01$, the chi square values would need to be ≥ 3.84 or 6.64 respectively, as can be seen from statistical tables.

Therefore in this case, there is no significant difference from the outcomes that would be expected by chance.

4.4.4 Analyzing Relationships Between Two Variables: Linear Correlation and Regression

The methods you are most likely to use to study the relationship between two variables are linear regression and linear correlation, both of which are parametric tests with certain requirements that must be met. Regression is used to describe how an independent variable (X) can cause changes in a dependent variable (Y), allowing us to predict the value of Y if we know the value of X. This is particularly useful when reading off Y values on a calibration graph (e.g. as part of a protein estimation or an enzyme assay). However, correlation simply helps to describe two variables that are associated with each other and vary together. It does not necessarily prove a cause and effect relationship between the two variables.

Simple linear regression helps to describe a causal or predictive relationship between two variables using a straight line graphical plot. The equation of the straight line is as follows:

$$\mathbf{y} = \alpha + \beta\mathbf{x}$$

In this equation "x" represents the independently measured variable (*explanatory* or *predictor* variable), whereas "y" is the dependent variable (i.e. the *response* variable), "α" is the intercept on the y axis (i.e. the mean value of y when $x = 0$) and "β" is the gradient of the line (i.e. the changes in y values per unit change in x).

However, for linear regression to be applicable, certain conditions have to be met:

Firstly, a straight line must be the true relationship between the two variables. Secondly, for any value of x, there must be a distribution of possible y values, the mean of each of which falls on the straight line, and the variance of which is equal at all values of x. Thirdly, the distribution of possible y values at each x should have a normal distribution. Finally, all observations must be made independently of each other. This assumption can be met only by appropriate experimental design.

As for other parametric tests, data transformation may be required to achieve a linear relationship. In general, however, conversion of data to $\log_{10}$ or natural log (ln) is most likely to correct a problem of non-linearity. To check linearity, a scatter plot between y and x values can be generated by hand or, more accurately, using computer packages such as Microsoft Excel or Minitab. There are several other ways to check normality of data value distributions such as a histogram (looking for a symmetrical profile about the mean) or a normality plot in statistical packages. If the relationship between y and x is clearly not linear, transformation of data may help but would be a pointless exercise in cases where the data distribution curve contains distinct peaks or valleys (e.g. a bimodal distribution). Transformation may work if a curve decreases or increases uniformly from left to right. Transformations can be made on either y or x (or both) values, whichever works best. The method adopted depends on the exact shape of the curve. For example, taking the log value of one variable may help to linearize

Figure 4.2 *Linear regression in a protein assay calibration curve. Shown is an example of a typical protein assay curve plotted using Microsoft Excel with linear regression applied. The assay used the bicinchoninic acid (BCA) reagent and various concentrations of protein standard (bovine serum albumin; BSA), as indicated. The absorbance of the protein following incubation with BCA reagent was recorded at a wavelength of 570 nm. In this case a good fit was observed with a high correlation coefficient (R^2), indicating a reliable standard curve. The equation of the straight line can be used to directly calculate the protein concentration value (x) corresponding to a given absorbance measurement (y).*

the data on an otherwise exponential curve. Obviously, the closer the correlation coefficient is to a value of 1 (typically 0.99) the better the evidence of a linear relationship between the two variables. An example of linear regression analysis and its presentation in graphical form is shown in Figure 4.2.

Most students by this stage will be familiar with the use of Microsoft Excel for basic graphical presentation. The package can also be used to apply a number of basic mathematical and statistical formulae (e.g. linear regression, descriptive statistics, t-tests, etc.) to aid data analysis. However, other packages exist that do a similar job.

A more powerful statistical package, (e.g. Minitab) is recommended for more in depth analysis, as it has a more comprehensive range of statistical tests and analyses, and a comprehensive help section. However, it may not be as effective as Excel or other software as a graphical package. As "cells" of data can be readily transferred from one package to the other, combined use of the two packages using their respective strengths seems like a reasonable compromise.

Information on using Minitab for data analysis is usually available in the "*Help*" facility on the statistical analysis package toolbar.

You should always remember that statistical analysis should be used to support your data by confirming its validity under rigorous and appropriate testing methods. It should not be used to try to get something out of a bad set of data – if the data are unreliable or inconsistent, it is better to repeat or redesign the experiment.

Uses of statistics

"If your experiment needs statistics you ought to have done a better experiment" – Ernest Rutherford

"Statistics are used much like a drunk uses a lamp post; for support, not illumination" – Vin Scully

4.5 Presentation, Structure and Organization of Data in Laboratory Reports

Effective presentation of your data is a key element in writing a good laboratory report. Developing these skills now will pay dividends later when you are analyzing more complex sets of data obtained from laboratory classes in later stages of your studies, from laboratory work on work placements or from your research project.

You will normally receive specific instructions on how to write the report from the lecturer in charge of the laboratory class. However, the following hints and tips may help you to produce a good report.

Structure

- Unless instructed otherwise by your tutor, a good report should typically be organized into sections containing an introduction, results and discussion/conclusions.
- Some reports may also require an abstract (summary), rationale, aims and/or a hypothesis, and a reference list citing sources of information used. Large reports may also benefit from a page indicating the main contents of the report and the page numbers where they can be found.
- Make sure that you include all of the sections requested in the assignment guidelines provided by your tutor.

Use of English

- A well written report helps to give a good impression to the person assessing it. Thus, if you are writing by hand, try to make your writing as neat (legible) as possible and avoid untidiness.
- If your report is word processed make sure that you check the spellings and consistency in font style and size.
- You will often come across the situation in which spell check does not recognize a scientific word and suggests an alternative. Do not automatically assume that spell check is right; it is possible that the word is not on the spelling database. You can of course add the word to the spell check dictionary but make sure that you have spelt it correctly before doing so; otherwise every subsequent use of the word will have the same error.
- It is convention in most areas of bioscience to use the past tense in the passive voice when writing a report.
- Examples of typical scientific and grammatical writing errors are shown in Table 4.1.

Table 4.1 Typical mistakes in written scientific English

Incorrect Phrases	Suitable Alternative Phrases
Add one gram of NaOH[a].	One gram of NaOH was added.
or	
We added one gram of NaOH.	
Then add 10 mL of reagent A[a].	Then, 10 mL of reagent A *were* added[b].
or	
Then we added 10 mL of reagent A.	
Bovine Serum Albumin (BSA)[c]	Bovine serum albumin (BSA)
Foetal Bovine Serum (FBS)[c]	Foetal bovine serum (FBS)
Sodium Chloride[c]	Sodium chloride
We homogenised the samples for...	The samples were homogenised for...
or	
Homogenise the samples[a] for...	
Samples were spun[d].	Samples were centrifuged.
Put samples in ependorfs[d].	Samples were placed in Eppendorf tubes
Samples were vortexed[d].	Samples were vortex mixed.
The experiment worked because we got positive results[e].	The experiment worked well as the expected results were obtained.
We got negative results[e] with compound A.	No effect was observed with compound A.
	or
	Compound A had no effect.

(*continued overleaf*)

95

Table 4.1 (*Continued*)

Incorrect Phrases	Suitable Alternative Phrases
Graph 1 shows…	Figure 1 shows…
We never saw any reaction.	No reaction was observed under the conditions specified.
Low temperature inhibits the enzyme a lot.	Enzyme activity was significantly reduced at lower temperatures.
I think this means…	This suggests…
	or
	This indicates…
The aims of this laboratory class are…	The aims of this laboratory class were…
A bacteria[f]	A bacterium
A mitochondria[f]	A mitochondrion
A phenomena[f]	A phenomenon
As you can see…	As can be seen…
It could of…	It could have…
It should of…	It should have…
I conclude that…	In conclusion…

[a] This is what you would see in a protocol but you need to convert it to the past passive when writing a methods section for a report or thesis.

[b] Note the verb agreement in this case; the subject in this case is '10 ml of reagent A'.

[c] Make sure that you use the upper and lower case letters correctly in chemical and reagent names. The examples shown assume that the sentence starts with this item. If it appears in the middle of the sentence, all words should be in lower case.

[d] These are typical examples of laboratory jargon used in general conversation. However a full description must be given in a report or thesis. For example, 'Eppendorf' is the name of a company which makes tubes, centrifuges, pipettes and a range of other products. It is often incorrectly spelt as in this example.

[e] Statements like positive and negative results are either very vague or entirely meaningless. You need to be more precise in the description of your data.

[f] Do not use the plural form of these words incorrectly. The same goes for other words such as medium.

4.5.1 Data Presentation

- Unless otherwise instructed by your tutor, graphs, histograms, schematic diagrams and images, and so on, should be referred to as Figure 1, 2, 3, and so on. Always remember to refer to the figure or table, and so on, at an appropriate point in the text of your report. For example, 'As shown in Figure 1, there was...'

- Each figure or table should have a clear descriptive heading. For example 'Figure 1: Effects of organophosphates on acetylcholinesterase activity' clearly indicates the nature of the experiment (i.e. an enzyme assay in the presence and absence of organophosphates) from which the data shown in the graph have been obtained.

- Do *not* refer to graphs as graph 1, graph 2, and so on, unless specifically requested to do so. Similarly, tables should be referred to as Tables 1 and 2, and so on, again with a clear descriptive title.

Figure 4.3 *Effects of toxin on neurite outgrowth (This is an example of a histogram plot with error bars). Neuronal cells were incubated with or without (CON) the concentrations of toxin indicated and the effects of toxin measured on the outgrowth of cellular extensions ('neurites'). The experiment was repeated four times and the mean ± SEM calculated for each experimental condition. The plot was using Microsoft Excel and incorporated error bars for SEM values as shown. Asterisks indicate where there was a significant of difference from the control (p < 0.05) using analysis of variance (ANOVA).*

Table 4.2 *The effects of toxin exposure on the morphology of cultured neuronal cells (This is an example of how to effectively annotate a table of data)*

Incubation Time	Treatment	Axons	Extensions	Round Cells	Flat Cells
4 h	CONTROL	32.35 ± 2.2	29 ± 1.1	16 ± 0.5	84 ± 0.5
	TOXIN	$13.2 \pm 1.4^*$	29 ± 1.1	15 ± 0.6	85 ± 0.6
24 h	CONTROL	35.9 ± 6.5	24 ± 0.8	19 ± 0.8	81 ± 0.8
	TOXIN	$18.5 \pm 1.1^*$	28 ± 1.5	25 ± 2.0	75 ± 2.0

Cultured cells were incubated in the absence (control) and presence of toxin for 4 and 24 h, after which their morphological features were assessed by light microscopy. Shown are the average counts $\pm$ SEM for four independent experiments. Asterisks indicate where changes were deemed to be significantly different from the control ($p < 0.05$; students' t test). The data suggest that the toxin has an inhibitory effect on the number of axons produced but no effect on other aspects of cell morphology.

- Another important feature of good data presentation is to have clearly annotated figures and tables. Thus, in a table each column and row should be clearly labelled to indicate what is being shown.
- Figures containing graphs need clear labels on both axes (e.g. Absorbance and concentration could be the labels on the y and x axis of a calibration graph for a protein assay as shown in the earlier example of linear regression).
- If a quantitative value is presented (e.g. activity or concentration) make sure that the units are correctly defined (refer to Chapters 1 and 2 for correct expression of units of size and concentration). Where possible give an indication of the spread of data. For example, if several replicate measurements were made in class, you could estimate the standard deviation and express it in the form of error bars in the figure or table.

Examples of appropriately annotated tables and figures are shown in Table 4.2, Figures 4.2 and 4.3. Hints and tips on presenting microscopy images have already been discussed in Chapter 2. Errors in any of the above factors will lose you marks.

4.5.2 Content

Introduction
- If required, this should give information about the background to the study, leading into the aims of the laboratory class.

- It may require you do some background reading in text books; if this is the case, sources of information should be cited in the text and listed in full at the end of the report.
- This section could involve a mixture of present and past tenses, as appropriate.

Aims

- Clear, concise bullet points indicating the main experimental objectives of the laboratory class.
- This is best done using the past tense (i.e. The aims of this laboratory class/ project were. . .)

Results

- Unless otherwise instructed, you should give a simple commentary on the trends observed in each experiment referring to each figure or table as appropriate.
- Present your tables and figures as indicated in the data presentation section above.
- Make sure that you refer to the appropriate tables and figures in your commentary.

Discussion

- In this section you should discuss the significance of the effects observed in your experiments (i.e. what does it *mean*). The depth required will vary from report to report, depending on the amount and complexity of the experiments involved.
- In longer reports (e.g. formal reports or project reports) you will be expected to relate your findings to what is already known (i.e. published), some of which you may have already covered in the introduction to the report. This is the opportunity to say whether or not your results agree with those reported by others and explain (discuss) any potential disagreements (e.g. on the basis of the experimental approach or materials used).
- Some mention of sources of error (e.g. sampling error, pipetting errors, inaccurate measurements, etc.) and possible ways in which the reproducibility of the data could be improved (e.g. performing more replicates and taking an average) would be a good idea in the section.
- Don't blame the protocol if everyone else got the experiment to work.

References

- Reference citation and reference list format should be consistent throughout.
- You should follow one recognized style. Numerous styles are used in books and journals that you will read; however, one of the most common formats is the Harvard system although there are numerous variations in use.
- The main emphasis should be on a *consistent* style, as this will look more professional. This system cites sources of information in the text by author surnames (if only one or two authors) or first author surname followed by *et al.*,

then the year of publication. For example, 'Smith (1980), Smith and Jones (2001) or Jones *et al.* (2009)'.

- An alternative, equally acceptable form of referencing is the numerical system, which cites references by placing numbers in parentheses or superscripts according to the order in which they appear in the text. The full reference list should be included in the same numerical order at the end of the report.
- NEVER put full details of the reference in the text.
- Full details of the reference source should then be given in the reference list at the end of the report. For example, an acceptable reference list format could be:

4.5.3   Articles in Journals

Daleo, G.R., Piras, M.M. and Piras, R. (1974) The presence of phospholipids and diglyceride kinase in microtubules from different tissues. *Biochem. Biophys. Res. Commun.* 61, 1043–1050.

4.5.4   Articles in Books

Hatch, R.C. (1988) Poisons causing nervous stimulation or depression. In: *Veterinary Pharmacology and Therapeutics* (Booth, N.H. and McDonald, L.E., eds.), pp. 1053–1101, Iowa State University Press, Ames, Iowa.

4.6   Summary

Take special care in the way that you analyze, interpret and present your data. There is no point in carrying out the technical aspects of laboratory work competently only to present the data obtained badly in your report.

In terms of data analysis:

- It is important that you perform appropriate statistical analysis, which will depend on the design of the experiment and the spread of the sample values obtained.
- If repeat measurements are made you should at least perform basic descriptive statistics.
- If sample means are being compared you need to determine the distribution and nature of the data sets before selecting an appropriate test.
- Statistical analysis should be used only to support your conclusions, always bearing in mind that the limiting factor in statistical analysis is the quality of the data being analyzed.

In terms of data presentation:

- Always follow the agreed assignment guidelines.
- Present your work in a neat and tidy manner using good scientific English.
- Write clearly and concisely, making sure that the appropriate sections are appropriately completed.
- Use clear headings, annotation and correct units on graphs and figures.
- If the experiment has not worked in your case, check the results of other students and discuss your own findings and what you should have expected, explaining possible reasons for the problems observed.
- If additional discussion is required, link your data to what is known in the literature and/or suggest possible sources of error and potential improvements.

Limitations of statistics

"I can prove anything by statistics except the truth." – George Canning

"The individual source of the statistics may easily be the weakest link" – Josiah Stamp.

5

THE EXTRACTION AND CLARIFICATION OF BIOLOGICAL MATERIAL

5.1 General Introduction

The basis of many biological experiments is the extraction and enrichment of cellular material. This usually requires two operational phases (a) the extraction process and (b) the clarification and enrichment of the component of interest. The method of extraction will depend upon the starting material; this could be abattoir material, plant tissue, eukaryotic or prokaryotic cell cultures. In addition to the choice of starting material, appropriate buffers must be used (possibly with additives) to ensure that the maximum amount of the desired component has been extracted.

5.2 Extraction

5.2.1 Proteins

Proteins are present within a cell at a high concentration and they reside in a reducing atmosphere (little if no contact with oxygen). An extraction process immediately alters the normal environment of cellular protein by diluting the protein in a liquid which is likely to be relatively oxygen rich due to the mechanical nature of the extraction process. As proteins are also heat and pH sensitive, the extraction of proteins from cells or tissue requires careful planning before the extraction process can be undertaken.

The method for the wholesale extraction of protein from cells or tissue will depend upon the starting material (animal, plant or bacteria).[1] In general a buffer (0.1–0.25 M) cooled to 4 °C with the pH adjusted to 7.5 is a good starting point. The addition of other reagents to improve the recovery can be beneficial. Reducing agents such as dithiothreitol (DTT) or 2-mercaptoethanol (2-ME) can be added at (1–5 mM) to help moderate the deleterious effects of the oxidizing extraction process. Chelating agents such as EDTA can be added to mop up any divalent heavy metal ions.

Basic Bioscience Laboratory Techniques: A Pocket Guide, First Edition.
Philip L.R. Bonner and Alan J. Hargreaves.
© 2011 John Wiley & Sons, Ltd. Published 2011 by John Wiley & Sons, Ltd.

Table 5.1 *The major classes of proteinases and inhibitors*

Class of Proteinase	Popular Inhibitor	Working Range
Serine	Phenylmethylsulphonyl fluoride (PMSF)	0.1–1 mM
Sulphydryl	Iodoacetamide	10–100 μM
Acidic	Pepstatin	1 μM
Metallo	EDTA	1–10 mM

In eukaryotic cells, hydrolytic enzymes are stored in the lysosomes of animal cells and the vacuole of plant cells. These compartmentalized enzymes do not normally come into contact with cellular proteins but during the extraction process the lysosomes (vacuole) will be broken and will mix with cellular proteins. There are two groups of hydrolytic enzymes that degrade proteins; proteinases (endoproteases) cleave the internal peptide bonds in proteins and peptidases (exopeptidases) sequentially clip off amino acids from a proteins amino or carboxy terminus. There are four major classes of proteinases (see Table 5.1) (a) serine proteinases possess a serine residue at the active site of the enzyme (b) sulphydryl proteinases possess a cysteine residue at the active site (c) acidic proteinases have a low pH optimum and (d) metallo proteinases have a requirement for metal ions. The action of these enzymes can be controlled by the addition of a proteinase inhibitor cocktail (see Table 5.1), thus improving the yield of intact proteins.

In addition to the additives outlined above, the extraction of protein from plant material can benefit from increasing the concentration of reducing agents up to 15 mM and including ascorbic acid (5 mM) to help moderate the action of endogenous oxidases, the action of which can cause protein aggregation and denaturation.

5.2.2 Deoxyribonucleic Acid (DNA)

DNA is typically extracted from cells or tissue by disrupting the cell membrane with high salt and EDTA. Sodium dodecyl sulphate (SDS) is then added to help unravel the structure of extracted proteins (see Chapter 6) and proteinase K (a non specific proteinase) is added to digest the protein. After an overnight incubation, the DNA present in the extract can be isolated by either (a) repeated washing in high salt (6 M NaCl) or (b) solvent precipitation (iso-propanol) and washing (ethanol). The purity and quantity of the DNA extracted can be assessed using spectrophotometry (see Section 3.8.1).

5.2.3 Lipids

Lipids can be extracted from cells or tissues with different combinations of solvents. Chloroform and methanol mixed in a ratio of 2:1 is a popular choice. When these solvents are mixed with an aqueous sample, two phases are generated. The chloroform rich lower layer will contain virtually all the lipids and the methanolic upper layer will contain the non lipid contaminants. Other solvent mixtures can be used including hexane/iso-propanol and ethyl acetate/ethanol. The total lipids extracted can then be separated into different fractions using thin layer or normal phase chromatography (see Sections 7.5 and 7.6).

KEY POINTS TO REMEMBER 5.1

When working with solvents it is best to use glass containers as some plastics may not be compatible with the solvents. Solvents can be hazardous to health and the environment so always work (and dispose of the used solvents) in accordance with the local health and safety procedures.

5.2.4 Organelles

The organelles of eukaryotic cells (e.g. mitochondria, chloroplasts, peroxisomes, Golgi apparatus) possess a lipid bilayer structure similar to the plasma membrane surrounding the cell. To obtain intact organelles, an isotonic buffered solution (0.2–0.25 M sucrose) should be used. This will prevent the organelles from shrinking in a hypertonic solution or expanding (possibly bursting) in a hypotonic solution. The organelles can be harvested by differential centrifugation (see Section 5.6.1) and possibly further purified on a density gradient (see Section 5.6.2).

5.3 Extraction Methods for Animal and Plant Tissue

There are many different extraction methods but to get the best yield from different starting tissue, certain techniques are more appropriate than others.

5.3.1 Mortar and Pestle

Small amounts of tissue can be conveniently extracted in buffer using a mortar and pestle.

- The mortar, pestle and buffer should be pre-cooled. Set the mortar on ice for the grinding process.
- Trim the fatty deposits from the tissue and cut the tissue into small pieces using a scalpel or scissors.
- Place the tissue into the mortar on ice with a small amount of buffer and using the pestle, grind the tissue into a paste.
- Add more cool buffer if necessary (a final tissue to buffer ratio of 1 : 2 (w/v) or 1 : 4 (w/v) is usually sufficient).
- Strain the extract through four layers of muslin to remove large particles and then clarify the extract using centrifugation (see Section 5.5).

Acid washed sand is sometimes added to the mortar to help in the extraction of plant material (or any other cells surrounded by a cell wall) as this can help disrupt the fibrous plant tissue. The processing of these extracts would be similar to the procedure outlined above.

As an alternative method, the tissue (animal or plant) can be frozen in liquid nitrogen (take appropriate health and safety precautions when using liquid nitrogen) and then ground into a powder prior to the addition of buffer.

5.3.2 Homogenizers and Tissue Grinders

As an alternative to a mortar and pestle, small amounts of animal tissue (or cells) can be disrupted using a homogenizer (see Figure 5.1). The pestle can be made from glass (Dounce) or from Teflon (Potter-Elvehjem). The pestle can be plunged and rotated by hand or by a motor. Heat generated during the extraction process can be dissipated by surrounding the glass mortar in ice. This method of extraction will also work for cultured plant cells but in general it is not abrasive enough for most plant tissue.

5.3.3 Blenders

Larger amounts of tissue whether animal or plant in origin can be extracted in a blender (e.g. Waring). The extraction vessel of a Waring blender (similar to a domestic blender) can be pre-cooled and small fragments of tissue can be added onto the cutting blades at the bottom of the vessel. Enough cooled buffer should be added so that a vortex is generated when the blender is switched on. The extraction time can vary (5–20 sec) but it should be interspersed with a cooling phase (place the blender on ice for 60–120 sec). After the blender has fragmented the tissue, the extract can be filtered through muslin (or cheesecloth) and then clarified by centrifugation (see Section 5.5).

Figure 5.1 *A picture of a hand held Potter-Elvehjem (glass-teflon) homogenizer. Taken from http://uk.vwr.com/app/catalog/Catalog?parent_class_id= 9&parent_class_cd=60478.*

A Polytron homogenizer (e.g. Ultraturrax®) is another type of blender with a rotating blade at the end of a hand held probe. The probe can be placed into a buffered solution on ice containing the tissue. When the probe is switched on, the tissue is drawn through narrow gaps at the tip of the probe towards the rotating blades, resulting in tissue fragmentation.

5.4 Extraction Methods for Bacteria

Bacterial cells (diameter $\approx 0.5\,\mu m$) are approximately the same size as eukaryotic mitochondria. They are surrounded by a peptidoglycan cell wall made from polymers of sugars and protein. The presence of the cell wall and the relatively small size of bacteria make them difficult to break open by the homogenization techniques used for the extraction of animal and plant tissue outlined above.

Small volumes of bacterial cultures collected by centrifugation ($13\,000 \times g$ for 20 min; see Section 5.5) can be broken open by incubating the bacterial suspension in a buffered solution of lysozyme.[2] This enzyme (also present in tears) weakens the cell walls of bacteria by hydrolyzing the bonds that hold the sugar polymers together. A detergent (e.g. SDS: sodium docecyl sulphate) to disrupt the bacterial plasma membrane may be included to ensure the complete disruption of the bacterial cells. Large volumes of bacterial cultures can be collected into a pellet by centrifugation (see Section 5.5) and diluted into an extraction buffer before being placed into the chamber of a pressure cell (a thick steel cylinder). The piston of the pressure cell is placed into the chamber on top of the bacterial cell suspension. The loaded pressure cell is placed into a mechanical press then subjected to pressure (10 000–40 000 psi). A valve at the base of the pressure cell allows the pressure to be released, forcing the bacterial cell suspension through a small aperture. The shattered bacteria in the extraction buffer can be collected into a beaker on ice. The majority of the bacteria will be fragmented in this process, which can be repeated to ensure complete extraction.

A sonicator provides a focused output of energy in the form of ultrasound waves (20 KHz) that can create holes in cell membranes, which results in cellular disruption. Sonicators can be used to fragment fragile cultured animal or plant cells but bacterial cells may also need the presence of small beads (Ballatini beads) to knock holes in the bacterial cell wall and plasma membrane.

KEY POINTS TO REMEMBER 5.2: EXTRACTION

- Different combinations of extraction methods may prove more effective than a single method.

- The extraction process will generate heat, which will denature proteins. Always pre-cool the extraction buffer and equipment (if possible) before use and extract on ice or include cooling cycles (e.g. 15 sec extraction followed by 30 sec on ice).

- Re-extracting the pellet after the first centrifugation can improve the yield.

- Check the pH (see Chapter 1) of the extract after extraction and re-adjust to the starting pH as necessary.

5.5 Clarification

After disrupting the tissue or cells of interest, a cell extract will contain a mix of soluble and non soluble material. In the extraction buffer there will be soluble cellular components mixed with clusters of non disrupted cells, different cellular organelles and/or fragments of cellular membranes and insoluble cellular components (e.g. starch from plant cells). The largest of these fragments and clumps of cells can be removed by filtering the extract through muslin (or cheesecloth) but the final clarification will require centrifugation.

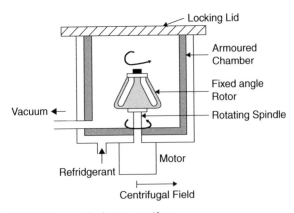

Figure 5.2 *A diagram of a basic centrifuge.*

5.5.1 Introduction to Centrifugation

In biosciences there is often a need to separate the item of interest from other components; for example, blood samples contain both red and white cells which can be separated from the blood plasma using centrifugation.

Any particle suspended in a liquid will sediment (or float) depending upon the size (and shape) of the particle, the viscosity of the medium and the density difference between the particle and the liquid. The rate of movement of the particle in a liquid is also influenced by gravity ($g = 981$ cm s^{-2}) with large dense particles sedimenting first. To separate particles in a liquid using these properties, it is often necessary to increase the gravitational force.

Gravitational force can be increased in a controlled environment using centrifugation in a machine called a centrifuge. Centrifuges are common instruments in any bioscience laboratory, providing a straightforward method for clarifying extracts. They produce a centrifugal force by spinning a sample about a central axis (see Figure 5.2). Depending on the type of centrifuge, the centrifugal force can be varied from a few hundred g to over half a million g by altering the rate of rotation. The relative centrifugal force (RCF) depends upon the speed of the rotor (see Section 5.5.3) in revolutions per minute (rpm) and the radius of the rotor (r; in mm). Modern centrifuges allow the running conditions to be set in relative centrifugal force (RCF). If the machine does not include this facility the data sheet for the rotor will show the average radius or a tape measure can be used as a last resort. Worked Example 5.1 (below) shows how to convert revolutions per minute into RCF. Alternatively, if the measurements are known, a nomogram (Figure 5.3) can be used to calculate the RCF.

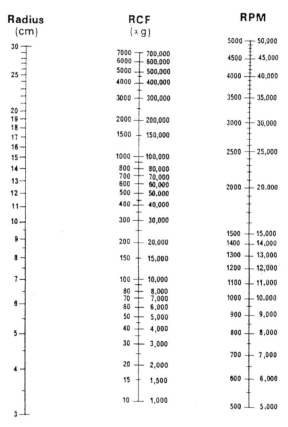

Figure 5.3 *A nomogram that can be used to calculate the RCF of a centrifuge rotor. To determine the maximum RCF measure the radius (mm) from the centre of the centrifuge rotor to the furthest point on the centrifuge tube as it lies in the rotor while spinning. Draw a line from the right hand column (radius value) to the left hand column (rotor speed in rpm). The RCF value is where the line crosses on the middle column.*

Worked Example 5.1: Conversion of rpm to RCF

Relative centrifugal force (RCF) = $1.118\, r\, (\text{rpm}/1000)^2$

rpm = revolutions per minute

r = the radius of the rotor in mm

- Calculate the RCF in a centrifuge rotor spinning at 3000 rpm if the average radius of the rotor is 95 mm.

$$
\begin{aligned}
\text{RCF} &= 1.118 \times 95 \times \left[\frac{3000}{1000}\right]^2 \\
&= 1.118 \times 95 \times (3)^2 \\
&= 956 \times g
\end{aligned}
$$

What would the RCF be if the speed of the same centrifuge were increased to 6000 rpm with the same average radius of the rotor (95 mm)?

$$
\begin{aligned}
\text{RCF} &= 1.118 \times 95 \times \left[\frac{6000}{1000}\right]^2 \\
&= 1.118 \times 95 \times (6)^2 \\
&= 3824 \times g
\end{aligned}
$$

Note: There is not a linear relationship between rpm and RCF.
In this example, the speed of the centrifuge in rpm has been doubled but the RCF has increased four fold. For this reason, it is accepted practise to quote the conditions for centrifugation in RCF using the average radius of the rotor rather than rpm. Information on the RCF, the temperature and the time of the centrifuge run allows the centrifugation conditions to be reproduced on another centrifuge using rotors with different diameters.

5.5.2 Centrifuges

Centrifuges (see Figure 5.2) are machines with an electrically driven central spindle, onto which a rotor is placed. For safety reasons, the spindle and rotor are contained within an armoured chamber with a locking lid. Centrifuges differ in their top speed, the size and volume of rotor they can accommodate and whether they are refrigerated or not.

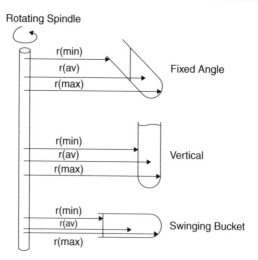

Figure 5.4 *The radius measurements in different centrifuge rotors. Minimum (r_{min}) average (r_{av}) and maximum (r_{max}) distances from the centre of the rotor.*

Micro-centrifuges (microfuges) are capable of rapidly accelerating small volumes (contained in 0.5, 1.5 or 2.0 ml plastic tubes; called microfuge tubes or Eppendorf tubes) to maximum speed of about 12 000 rpm (RCF 10 000 × g).

Bench-top centrifuges or low-speed centrifuges can be used to harvest cells or large particles. They usually have a maximum speed of 3000–6000 rpm. The maximum RCF value will vary depending on the rotor used.

High-speed centrifuges are larger machines with a top speed of approximately 25 000 rpm (depending on the rotor, the RCF can be up to 75 000 × g). They are used to harvest protein precipitates and larger sub-cellular organelles. These machines need to be refrigerated to counter the frictional heat generated as air passes over the rotating surface of the rotor during the course of the run. There are also bench top versions of this type of centrifuge which are useful for smaller volumes of sample.

Ultracentrifuges are specialized machines that can operate at speeds up to 80 000 rpm (depending on the rotor the RCF can be up to 350 000 × g). Slow acceleration and deceleration rates are usually available to prevent sample mixing when in "swinging bucket" rotors (see Figure 5.4). Ultracentrifuges require the chamber to be evacuated and refrigerated before the run can commence to higher speeds. They are used to sediment cellular membranes as well as DNA and RNA. Bench-top ultracentrifuges can be used to subject small sample volumes (0.2–5.0 ml) up to 650 000 × g.

Continuous (flow) centrifuges can be used for harvesting cells from large volumes of media as the liquid flows continuously through a special rotor.

5.5.3 Centrifuge Rotors

Centrifuge rotors can be divided into three different types (a) fixed angle rotor (b) swinging bucket rotor and (c) vertical rotor (see Figure 5.4). These different types of rotors are selected depending upon (a) the type of centrifugation required (differential, rate zonal or isopynic) (b) the volume of sample available and (c) the RCF required to complete the experiment.

Micro centrifuges and analytical centrifuges are usually fitted with fixed angle rotors engineered from a solid block of aluminium, titanium or carbon fibre. These rotors are a popular choice for routine clarification tasks. The rotors must have a smooth surface to reduce wind resistance and be conductive to dissipate the heat generated by movement through air at high speeds. Routine checks for surface scratches or metal fatigue should be undertaken on a regular basis.

When a sample requires pelleting in a fixed angle rotor, the particles in the liquid move away from the centre of rotation throughout the whole centrifuge tube. They eventually contact the wall of the tube (see Figure 5.5b) and then start to slide down the wall of the tube to the furthest distance from the centre of rotation forming a pellet. If the run time of the centrifugation is sufficient, the pellet will be located at r_{max}. Care should be taken when removing a sample after centrifugation in a fixed angle rotor to avoid disturbing the precipitated pellet.

Fixed angle rotors come in a variety of different sizes but large rotors capable of accommodating larger volumes have a lower maximum speed limit and hence a lower maximum RCF. The K factor of a rotor (supplied in the rotor data sheet) determines the pelleting efficiency of a rotor (a low K factor indicates a higher pelleting efficiency) and this can be a useful parameter when comparing the use of different rotors.

Bench top centrifuges are routinely fitted with swinging bucket rotors. These swinging bucket rotors are useful for small and large volume of sample (10–1000 ml) for example the harvesting of cells from culture or processing blood samples. High speed centrifuges also have swinging bucket rotors for rate zonal or isopynic centrifugation (up to 36 ml) (see Figure 5.5a and Section 5.6.2 and 5.6.3).

The samples are loaded into buckets in the vertical position and, during the course of a centrifugation run, the buckets move into the horizontal position. In a differential centrifugation experiment, all the particles in the liquid will be subjected to maximum RCF (see Figure 5.5b) and they will eventually pellet at the bottom of the centrifuge tube.

Figure 5.5 *What happens to particles during the course of a centrifugation run (a) gradient conditions (b) pelleting conditions.*

As vertical rotors have specialized applications they are not commonly used. However, they are useful in isopycnic separations (see Figure 5.5a and Section 5.6.3) of DNA in caesium chloride gradients.

KEY POINTS TO REMEMBER 5.3

- Regular maintenance of a rotor will prolong its life and help prevent accidents.

- After a centrifugation run has finished, always rinse the rotor with water to remove any spillages and then leave the rotor upside down on tissue to dry.
- Dislodge any debris with a *soft* cloth or tissue, not a brush.

5.5.4 Centrifuge Tubes

There are many different types of centrifuge tubes manufactured from a variety of materials to accommodate diverse volumes. Before selecting a centrifuge tube, consult the data sheets from the manufacturers on the most appropriate tube to use and their chemical compatibility. Regular glass tubes should not be used above $3000 \times g$. Toughened Cortex™ glass can be used up to $25\,000 \times g$, whereas polycarbonate (clear), polyethylene (clear), polypropylene (opaque) and polyallomer (opaque) can be used up to $100\,000 \times g$. Thicker walled tubes may be required for speeds above $100\,000 \times g$.

KEY POINTS TO REMEMBER 5.4

- Centrifuge tubes are used to contain the sample during a centrifugation run. A failure of a tube's integrity during a centrifuge run could result in the rotor becoming unbalanced, damaging the rotor spindle. Furthermore, hazardous or corrosive material could be released into the rotor and the centrifuge chamber.
- If a tube shows signs of "wear and tear" it should be discarded.

5.6 Centrifugation Techniques

In the 1920s T. Svedberg and J.W. Williams undertook pioneering work in centrifugation. In honour of that work, the sedimentation velocity of a molecule (or organelle) in a centrifuge is called a Svedberg constant (sedimentation coefficient) (S). The Svedberg constant (1×10^{-13} seconds) is a ratio of a particle's sedimentation velocity (ms^{-1}) to the applied acceleration. Larger molecules sediment quicker and have larger S numbers for example eukaryotic ribosomes (80S), comprising a large (60S) and a small (40S) subunit. There are a number of different centrifugation techniques that can be used to clarify homogenates and enrich organelle preparations.

5.6.1 Differential Centrifugation

This form of centrifugation exploits the size and mass rather than the density of a particle to effect a separation. The pelleting of precipitates and the separation of eukaryotic organelles can be achieved using progressive increases in the RCF and the time this force is applied to particles. The different particles present will sediment in order of decreasing size and mass.

- A eukaryotic cell homogenate contains organelles with similar densities but there are differences in their size (see Table 5.2). Differential centrifugation can be used to separate these particles by a step wise increase in the RCF and the time that the RCF is applied to the extract (see Figure 5.6).
- The cell aggregates, individual whole cells and nuclei can be collected in a pellet by centrifugation at $2000 \times g$ for 2–5 min.
- The supernatant from this first step can be centrifuged at $20\,000 \times g$ for 30 min to pellet the mitochondria and a mixture of membranes from the lysosomes and endoplasmic reticulum.
- The supernatant from this step can then be centrifuged at $80\,000 \times g$ for 60 min to produce a pellet of mixed microsomal membranes consisting of membranes from the plasma membrane, the lysosomes, the endoplasmic reticulum and the Golgi body (also other organelles such as the peroxisomes).
- The supernatant can be subjected to centrifugation at $150\,000 \times g$ for 180 min to pellet the ribosomes.

Figure 5.6 *Differential centrifugation of a mammalian tissue extract.*

116

Table 5.2 *The diameter and density of some eukaryotic organelles. (Based on Table 2.6 "protein purification" P. Bonner, Taylor and Francis (2007). p. 44)*

Organelle	Diameter (μm)	Density (g cm^{-3})	marker enzymes	Marker proteins
Nuclei	5–10	1.4	DNA dependent RNA polymerase	Lamin A
Mitochondria	1–2	1.19	Succinate dehydrogenase	Monoamine oxidase or cytochrome C
Lysosomes	0.2–2	1.12	Acid phosphatase	Lysosomal associated membrane protein (LAMP 1 and 2)
Endoplasmic reticulum	(vesicles of variable diameter)	1.13–1.15	Glucose 6-phosphatase	Calreticulin
Golgi	(vesicles of variable diameter)	1.03–1.06	NADP$^+$ phosphatase	ATP binding cassette transporter 1 (ABCA-1)
Plasma membrane	(vesicles of variable diameter)	1.02–1.04	Na$^+$ K$^+$ ATPase	Sodium potassium ATPase

The fractions produced at each step will not be pure, but they can be said to be enriched in a particular organelle. The purity of each fraction can be assessed by measuring the activity of enzymes (marker enzymes; see Table 5.2) known to be present only in a particular organelle. Alternatively, the proteins in each fraction can be separated by electrophoresis under denaturing conditions and transferred onto nitrocellulose during Western blotting (see Chapter 6). The blot can then be probed with antibodies to proteins from different organelles (see Table 5.2) to demonstrate enrichment or contamination.

5.6.2 Density Gradient Centrifugation

- Differential centrifugation of a eukaryotic cell extract will produce fractions enriched in a particular organelle. These fractions can be purified further by density gradient centrifugation technique.
- A step gradient of sucrose (or other reagents such as Ficoll, Percoll or Nycodenz) can be generated by layering progressively less dense sucrose solutions on top of each other (10–40% w/v).
- The fraction to be further purified is layered onto the sucrose step gradient and centrifuged in a swinging bucket rotor at $80\,000 \times g$ for 60 min.
- The membranes will migrate through each fraction until they are prevented from moving further by the density of the next layer of sucrose.
- At the end of the experiment the interfaces of each layer are enriched in membranes of a similar density and these can be removed by careful pipetting.

5.6.3 Isopynic Centrifugation (Equilibrium Density Gradient Centrifugation)

- As an alternative to density gradient centrifugation, the sample after differential centrifugation can be layered onto a continuous 20–70% (w/v) gradient of sucrose (or other reagents described above).
- During centrifugation, the components will migrate down the gradient, until they reach a point in the gradient where their density matches that of the sucrose.
- At the end of the experiment the layers can be removed by careful pipetting or by puncturing the tubes and collecting fractions.
- Vertical rotors are the most appropriate rotor for this technique but swinging bucket rotors can be used.

5.7 Points of Good Practice in Centrifugation

Prevention and regular maintenance is the best way to manage the risk of leaks. The correct use of the tubes, rotors and centrifuges will also help to prevent unnecessary damage.

- *Never* use a centrifuge unless you have received the appropriate safety training.
- *Never* overide any of the centrifuge's safety features.
- Log the use of the machine and provide information on the nature of the sample and how the operator can be contacted.
- Always check that the rotor is clean before use (rinse and clean if required; see Section 5.5.3).
- Switch the centrifuge on and place the rotor onto the spindle in the centrifuge bowl. Allow them both to cool to the operating temperature (the rotor may be pre-cooled in a fridge before use) before loading the samples.
- Only use the appropriate tubes for the intended rotor (see Section 5.5.4). Fill the tubes with distilled water and check for leaks prior to loading the samples.
- If the centrifuge tubes have been on ice, dry the tubes with tissue before placing them on the balance.
- *Never* add too much sample to the centrifuge tubes. Always fill the tubes to the level recommended by the manufacturers.
- Use a top pan balance to correctly balance pairs of tubes and remember to include the lids on the balance if they are to be used.
- Situate the balanced pairs of tubes in the rotor exactly *opposite* each other.
- When all the tubes have been loaded into the rotor, secure the lid of the rotor.[3] Close the lid of the centrifuge bowl (a locking sound is usually heard). If you are conducting a run in a high speed analytical centrifuge the vacuum pump will start.
- Programme the centrifuge according to the manufacturer's instructions and press the start button.
- It is advisable to stay with the machine until the required RCF (or rpm) has been reached. If the programme is halted and the speed returns to zero, open the bowl and check that there is no leakage into the rotor. If a leak has occurred clean the spillage in the bowl and rotor. Change the damaged tube and rebalance.
- When the centrifuge is operating at the required RCF (or rpm) value, it can be left until the run time has been completed.
- If the samples have been loaded into a fixed angle rotor, the precipitate will be at the furthest point from the centre of the rotating spindle. Carefully remove the centrifuge tube and try not to disturb the pellet.

- If the pellet mixes with the supernatant re-centrifuge the samples for a longer period of time at the same RCF or for the same period of time with an increased RCF value.
- After the centrifugation run has finished, *always* rinse the rotor with water to remove any spillages and then dry on tissue. Dislodge any debris with a *soft* cloth.
- To comply with health and safety procedures (and to ensure the machine's longevity), spillages must be reported and removed prior to the next centrifugation run.

5.8 Summary

- The cellular components of biological samples can be extracted by a number of different methods.
- To improve the yield of the component of interest, attention must be paid to the temperature, pH and components of the extraction liquid.
- Extracts can be clarified by increasing the relative centrifugal force (RCF).
- There are many different types of centrifuges for different applications
- In differential centrifugation the cellular components can be enriched by progressively increasing the RCF and the duration of the centrifuge run.

Notes

[1] If a specific protein is to be extracted, the conditions necessary to ensure good yields of this protein can be determined in small scale experiments.

[2] Another enzyme called lyticase can be used to disrupt yeast cells.

[3] Some models of micro centrifuges do not have lids and some models are fitted with a lid to prevent excessive noise.

6

ELECTROPHORESIS OF PROTEINS AND NUCLEIC ACIDS

6.1 General Introduction

Electrophoresis is a technique that is used to separate charged molecules in an inert support matrix under the influence of an electric field. When an electric field is applied, different biological molecules will move at different rates and the components of a mixture can be separated from each other. Electrophoresis has numerous applications in bioscience research, including the separation of proteins and nucleic acids, which will be the main focus of this chapter. It is typically used in the analysis of complex mixture (e.g. cell and tissue extracts) or in the monitoring of purity of isolated macromolecules (e.g. proteins, DNA, and so on).

The electrophoretic separation of any molecule depends on opposing forces of propulsion and resistance, as outlined below:

Propulsion is driven by:
a. Charge – negatively charged molecules (anions) migrate towards the anode (+) and positively charged molecules (cations) migrate towards the cathode (−).
b. Field strength – Higher voltage promotes faster migration up to a certain limit when overheating can occur, which will distort the electrophoretic migration pattern and could damage the resultant gel.

Resistance to migration is dependent on:
a. The size and shape of the biomolecules being separated –larger molecules migrate through a solution or matrix at a slower rate than smaller (lower relative mass) molecules as they experience greater frictional drag. Similarly, bulky molecules will experience more resistance than molecules with a linear shape.
b. The size of the pores in the support matrix – smaller pores will retard the rate of migration.
c. The viscosity of the buffer can slow the progress of molecules in electrophoresis.

Basic Bioscience Laboratory Techniques: A Pocket Guide, First Edition.
Philip L.R. Bonner and Alan J. Hargreaves.
© 2011 John Wiley & Sons, Ltd. Published 2011 by John Wiley & Sons, Ltd.

Thus, the electrophoretic migration of individual biomolecules is dependent on the balance of the propulsion and resistance forces that they experience.

6.2 Separation of Protein Mixtures by Gel Electrophoresis

The most commonly used method for separation of proteins is polyacrylamide gel electrophoresis (PAGE) in the presence of sodium dodecyl sulphate (SDS) commonly known as SDS-PAGE. This technique is typically used to estimate the molecular weight of a protein by comparison of its electrophoretic migration with that of protein standards of known molecular weight which are used to construct a calibration graph (discussed in Section 6.2.3). The key steps in sample preparation, electrophoresis, staining and data analysis for SDS-PAGE and a brief overview of other application of electrophoresis in protein separation are outlined below.

Note: Electrophoresis involves the use of reagents that are potential carcinogens and teratogens. For example, the acrylamide, which is used to form the separation gel, is highly toxic in its monomer form either in solution or as a powder. Always make sure that you handle and dispose of such reagents according to local health and safety guidelines.

6.2.1 Preparation of Electrophoresis Gels

Proteins are normally separated electrophoretically in a 3-dimensional cross-linked gel matrix formed from the polymerization of monomeric acrylamide in the presence of the cross-linking molecule *bis*-acrylamide. Typical ratios of acrylamide to *bis*-acrylamide are 29 : 1 or 37·5 : 1 and attention should be given to the fact that proteins will separate differently if changing from one cross-linker ratio to another in gel preparation. Any given protein might be expected to migrate more quickly at the lower cross-linker ratio (37·5 : 1) than at the higher one (29 : 1).

Nowadays, acrylamide is typically supplied as a prepared stock solution containing a mixture of acrylamide and *bis*-acrylamide in aqueous solution of either 30 or 40% (w/v) total acrylamide. This stock solution and distilled water are then mixed in various proportions with fixed amounts of Tris buffer and SDS, together with catalysts, to produce resolving gels of the required concentration for best resolution of the protein mixture to be separated. Typical recipes are shown in Table 6.1 for the preparation of resolving gels containing 7.5, 10 or 12% (w/v) polyacrylamide. Although a single resolving gel system can be used, better

Table 6.1 Composition of resolving gels and stacking gels for SDS-PAGE

	Polyacrylamide concentration (w/v)* all volumes are in ml		
(a) Resolving gel (10 ml)	7.5%	10%	12%
Effective range of resolution (kDa)	30–300	15–150	5–80
40% (w/v) acrylamide stock solution (29 : 1) (comprising acrylamide and N, N'-methylene bis-acrylamide in a mass ratio of 37.5 : 1 or 29 : 1)	1.9	2.5	3.0
1.5 M Tris buffer, pH 8.8	2.5	2.5	2.5
10% (w/v) SDS	0.1	0.1	0.1
Distilled water	5.5	4.9	4.4
Mix thoroughly then add catalysts in sequence:			
10% (w/v) ammonium persulphate (AMPS)	0.10	0.10	0.10
N, N, N', N'-tetramethylethylene diamine (TEMED)	0.01	0.01	0.01
Mix well, pour and overlay with water or solvent until polymerised			
(b) Stacking gel (10 ml)	4%		
40% (w/v) acrylamide stock solution (29 : 1)	1.0		
0.5 M Tris pH 6.8	2.5		
10% (w/v) SDS	0.1		
Distilled water	6.4		
Mix thoroughly then add catalysts in sequence:			
AMPS	0.10		
TEMED	0.04		
Mix well, pour onto the resolving gel surface and insert comb to form sample wells.			

resolution of protein bands is achieved using a resolving gel (for protein separation) overlaid with a stacking gel (which allows rapid entry and accumulation of proteins at the resolving gel interface).

Manipulation of the polyacrylamide concentration determines the effective molecular weight separation range, over which good resolution can be obtained. The amounts of water and acrylamide solution would need to be adjusted if changing from one acrylamide stock solution to another in order to get the same final concentration of acrylamide.

After mixing the main ingredients the initiator (ammonium persulphate) and catalyst (TEMED) are added and immediately mixed. The mixture is then poured in between two glass plates separated by spacers (1 to 2 mm wide), sealed at the bottom and mounted on a gel assembly stand (Figure 6.1). The mixture is overlaid with distilled water or water saturated solvent (e.g. isobutanol) to provide a flat surface for the gel by removing the meniscus. The overlay solution also prevents direct contact with air, which inhibits polymerization.

After polymerization of the resolving gel, which may take 1–2 h, the iso-butanol (or water) layer is removed, the resolving gel surface rinsed with distilled water and the stacking gel mixed and applied. Once again, the catalysts are added and thoroughly mixed into the acrylamide solution just prior to pouring it into the gel assembly unit. As can be seen for the recipe in Table 6.1, the stacking is at a lower pH (6.8) and the polyacrylamide concentration (4% w/v) is lower than that of the resolving gel. This facilitates the rapid entry of all proteins into the gel as electrophoresis commences and their subsequent concentration at the interface between the stacking and resolving gels prior to entry into the latter. The wells of the stacking gel are produced by the insertion into the freshly poured stacking gel mixture of a sample well 'comb', the teeth of which form the mould that shapes the wells into which samples are loaded (Figure 6.1). If you are pouring your own gels, a stacking gel depth of about 1 cm is ideal. If you are attending an SDS-PAGE practical class of 3 h or less, you may be provided with a pre-cast polyacrylamide gel. In this case you only need worry about following the instructions to open it, to remove the comb carefully and to rinse out the sample wells with running buffer as for the homemade gels.

The prepared gel assembly containing the homemade or pre cast gel matrix is inserted into the gel holding assembly with the smaller of the two glass plates facing inwards. The exact way in which this is done will vary slightly from one make of apparatus to another, but it basically involves careful positioning of the shorter plate of the gel assembly (or the appropriate surface of the buffer dam) against a rubber sealing gasket on the electrode unit so that they can be easily snapped into position. Either another gel or a 'buffer dam' (i.e. a piece of plastic shaped like the polymerized gel assembly) is applied to the other side of the electrode unit. This will create a water tight sealed inner chamber. Once

Figure 6.1 *Making a polyacrylamide gel. The type of gel most commonly used for protein separations is a dual gel system in which the resolving gel (which separates the proteins by size) is overlaid with a stacking gel (which allows proteins to enter the gel and accumulate at the interface with the resolving gel) of lower pH and acrylamide concentration. Shown are schematic representations of the major steps in gel formation. The stacking gel can be omitted but the resolution and sharpness of polypeptide bands is much better when a stacking gel is used. In such cases the gel mould is filled with resolving gel mixture and the comb applied directly into it.*

assembled with a gel on both side the inner chamber is filled to a level below the top of the larger glass plate but above the top of the lower glass plate of the gel assembly. Two typical types of apparatus are shown schematically in Figure 6.2.

6.2.2 Sample Preparation and Loading

In SDS-PAGE proteins are separated on the basis of size differences. However, as discussed above, the electrophoretic migration of a molecule depends on several factors including its size, shape and charge. As each protein is unique in terms of its primary amino acid sequence, it follows that the net charge will vary from protein to protein depending on the balance of amino acids with positively (e.g. lysine, histidine and arginine) and negatively charged amino acid (aspartic and glutamic acid) side chains. In order to eliminate such charge differences, prior to their separation by SDS-PAGE, proteins are denatured (i.e. unfolded) by boiling them in sample preparation buffer (Table 6.2) containing the anionic detergent sodium dodecyl sulphate (SDS). Like other detergents, SDS is an amphipathic molecule, meaning that it has hydrophilic and hydrophobic components. In this case the hydrophobic components (the alky chain) interacts non-covalently with hydrophobic amino acid side chains (e.g. valine), thus preventing the protein from precipitating in solution, whereas the hydrophilic portion (i.e. the sulphate group) projects outwards from the protein surface into the aqueous environment keeping the detergent covered polypeptide in solution.

As the SDS binds uniformly along the polypeptide chains, at approximately every two amino acids, each protein will now have a constant charge: mass ratio. However, as many proteins also contain disulphide bonds between cysteine residues their tertiary structure is not completely linearized unless a reducing agent is included in the sample preparation buffer. A common reagent used in this case is 2-mercaptoethanol, which breaks down disulphide bridges, thus allowing the complete unfolding and linearization of the proteins (Table 6.2). Proteins treated in this way become linearized polypeptide chains with constant charge: mass ratio, varying only in their chain length (i.e. mass).

After the stacking gel has polymerized (30–60 min), the sample well comb is carefully removed and the sample wells rinsed out with electrophoresis

◄ ───

Figure 6.2 *Schematic diagrams of typical apparatus lay out for SDS-PAGE. Shown are two of the major types of electrophoresis tank in cross section. The system in the top diagram employs an upper and lower buffer reservoir system with the cathode at the top. Some systems of this type allow a second gel to be attached to the other side of the upper buffer reservoir, with a lower buffer reservoir on the opposite side to the one shown. The system in the lower diagram employs an inner and outer buffer tank system with the electrodes arranged in such a way that protein migration is in the downward direction. The inset (top left) shows the alignment of the gel assembly with the rubber seal to create a water tight contact that prevents leakage. Indicated are the longer outer glass plate (a), the gel (b), the shorter inner glass plate (c) and the rubber seal (d). A similar type of seal is used in all systems. Note that only one terminal is viewed in the cross sections but a positive and negative terminal are present in both types of apparatus and these normally plug into sockets in the protective lids, connecting to the leads that are plugged into the power unit. Relatively few systems use horizontal (flat bed) gel electrophoresis, which is more common for nucleic acid electrophoresis (see example in Figure 6.5).*

Table 6.2 *Examples of typical electrophoresis buffer compositions*

Technique	Sample preparation/loading buffer	Electrophoresis running buffer
SDS-PAGE	2% (w/v) SDS in 62.5 mM Tris-base (pH 6.8), 10% (w/v) glycerol, 5% (v/v) 2-mercaptoethanol, 0.1% (w/v) bromophenol blue	0.1% (w/v) SDS in 25 mM Tris-base (pH 8.3) and 192 mM glycine.
Non-denaturing PAGE	As for SDS PAGE except that SDS and 2-mercaptoethanol are excluded	25 mM Tris-base (pH 8.3) and 192 mM glycine.
Non-denaturing DNA electrophoresis	4% (w/v) sucrose in TBE to which one or more tracker dyes (e.g. bromophenol blue) can be added at 0.01% (w/v) or lower concentrations.	90 mM Tris-base (pH 8.3), 80 mM boric acid, 2.6 mM EDTA (TBE) *Acetate can be used in place of borate to make 'TAE' buffer. TAE.
Denaturing DNA electrophoresis	50 mM NaOH, 1 mM EDTA, 2.5% (w/v) Ficoll (type 400), 0.01% (w/v) bromocresol green.	

Notes: Gel electrophoresis buffers are often prepared as master stocks (e.g. ×5 or ×10 strength) that are diluted to working strength with distilled water. The buffer compositions given are typical examples and represent typical (but not exclusive) working concentrations. You should also be aware that several variations exist, as indicated below:

In SDS-PAGE sample buffer, alternative reducing agent such as dithiothreitol can be used. Furthermore, in specific applications, a different pH may be used for both sample and running buffer.

In nucleic acid separations, sucrose and Ficoll are interchangeable and can be replaced by a similar concentration of glycerol. Various tracker dyes can be used and sometimes more than one dye is added to the sample buffer giving distinct dye fronts reflecting their different molecular sizes. Other denaturing agents may replace NaOH (e.g. urea).

running buffer, typically Tris, glycine and SDS (Table 6.2). Glycerol (10% w/v) is added to the sample buffer to make it more dense than the electrophoresis running buffer and a tracker dye (bromophenol blue) makes it more visible; thus, when samples are loaded directly through the running buffer using automatic pipettes they can be seen sinking to the bottom of the wells (see the practical hints and tips box for the best way to load your samples). It is good practice to apply a set of protein standards of known molecular weight with every gel. If the standards are pre-stained (i.e. they have a covalently bound dye attached to them) their migration can be monitored through the gel as electrophoresis proceeds.

6.2.3   Running Electrophoresis Gels

As all proteins after the sample preparation step are now uniformly negatively charged, when the electric field is applied, they will move towards the anode (+). The larger proteins (i.e. longer polypeptide chains) will move more slowly than the smaller ones, as they experience greater levels of frictional drag as they move through the gel. Thus, smaller proteins will migrate more quickly and arrive at the end of the gel first. A negatively charged tracker dye, usually bromophenol blue, is added to the electrophoresis sample buffer. The blue dye moves ahead of all proteins forming the dye front. When this reaches the end of the gel, the voltage is switched off and the gel removed carefully to avoid damaging the gel or the glass plates. Having dismantled the gel assembly, the gel is carefully lifted or floated off the plates into a staining solution that fixes and stains the proteins separated in the gel so that they can be visualized and an image taken.

Proteins can be revealed by staining with Coomassie brilliant blue in a mixture of acetic acid (5–10% v/v) and methanol or ethanol (10–20% v/v), which can detect down to 0.1–0.2 μg of protein. A typical stained gel is shown in Figure 6.3. A more sensitive staining reagent containing fluorescent protein binding dyes (e.g. Sypro® Ruby) or silver chloride can be used to detect as little as 0.5 ng of protein. Alternatively, there are a number of recently developed eco friendly Coomassie based dyes that are very rapid, reduce the use of solvents in a fixation step and have a greater degree of sensitivity (e.g. CooBlue MAX or Bio-Safe Coomassie stain); these are colloidal Coomassie blue solutions that can detect as little as 10–20 ng of protein within 1 h).

After staining, separated proteins appear as coloured bands of the same width as the sample wells into which they were applied. Once the gels are stained, it is

Figure 6.3 *A typical Coomassie blue stained SDS-PAGE gel. Proteins were separated by SDS-PAGE in a 7.5% polyacrylamide gel overlaid with a 4% polyacrylamide stacking gel. Shown, from left to right, are molecular weight standards (Std) and a protein extract at 10 μg (a), 3 μg (b) and 1 μg loadings. This extract was a preparation of microtubules purified from mammalian brain. The major protein (55 kDa) is tubulin and there are a number of less abundant microtubule proteins. It can be clearly seen that the staining is quantitative over this protein loading range (i.e. the staining intensity is proportional to the amount of protein present). The molecular weights of the protein standards are shown to the left of the gel image. Note that although silver staining is more sensitive than staining with Coomassie blue, it is less quantitative. An example of a silver stained gel is shown in Figure 6.4.*

possible to measure the electrophoretic migration distance of each polypeptide band and compare them to the distance moved by the dye front. An Rf value (retention factor or 'relative to the front') can then be calculated for each band, where:

$$Rf = \frac{\text{Distance moved by polypeptide}}{\text{Distance moved by the dye front}}$$

This will always be a ratio with no units and it will never be a value greater than one, as no protein molecule moves faster than the dye front. If a set of protein standards of known molecular weight are electrophoresed in parallel to the samples a calibration graph can be constructed between the $\log_{10}$ value of their molecular weight and their Rf values. This will be a linear negative slope, as indicated in the worked example below.

WORKED EXAMPLE 6.1: MOLECULAR WEIGHT DETERMINATION BY SDS-PAGE

Complete the following tasks:

a. Measure the Rf values of the standard protein and sample bands in the schematic diagram of an SDS-PAGE gel provided.
b. Plot a graph between the $\log_{10}$molecular weight (MW) and the corresponding Rf value for each standard.
c. Use the calibration graph to estimate the molecular weight of the sample band.

What to do:

a. To calculate the Rf value, you should measure the distances migrated from the start of the resolving gel (origin) to the centre of each band and divide this by the distance moved by the dye front (DF).
b. Put all of these values into a clearly annotated table as shown.
c. From the molecular weights given for the standards, work out the $\log_{10}$ values and put this in the corresponding column of the table, as shown.
d. It is then possible to plot a graph there by hand or using a software package such as Microsoft Excel, as shown below the table.
e. If plotting by hand, remember to stretch the log axis as much as possible. For example, as the log values range only between 4.45 and 5.31, the start and end of the x axis could be log 4 and 6, respectively (or even 4.4 and 5.4, as shown on the Excel plot). You should then draw the best fit straight line through the data points.
f. The advantage of using a package such as Excel is that the trend line can be fitted, and values for the equation for the straight line and the correlation coefficient can be displayed.

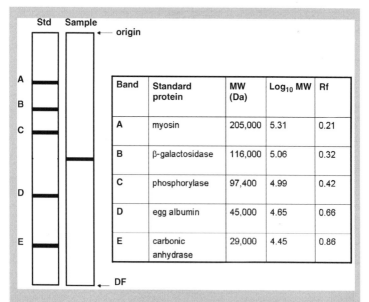

Band	Standard protein	MW (Da)	Log$_{10}$ MW	Rf
A	myosin	205,000	5.31	0.21
B	β-galactosidase	116,000	5.06	0.32
C	phosphorylase	97,400	4.99	0.42
D	egg albumin	45,000	4.65	0.66
E	carbonic anhydrase	29,000	4.45	0.86

Calibration graph for molecular weight determination by SDS-PAGE

$y = -0.7632x + 4.2283$
$R^2 = 0.9853$

The high value for the correlation coefficient (>0.98) indicates that the linear plot is good (See Chapter 4).

Now, do the calculation:

$$\frac{\text{Distance moved by the sample band}}{\text{Distance moved by the dye front}}$$

You should find a value of approximately **0.52**

Remember it is a *ratio* and therefore has no units! This value can then be read off on your hand drawn graph or substituted into the equation from the Excel plot shown above, giving a log10 molecular weight value of 4.858 884.

If you had drawn exactly the same graph by hand, the nearest you would get to this would be *4.86* (2 decimal places). Hand measurements cannot be more accurate than this.

You then need to calculate the antilog of this value in order to arrive at the estimated molecular weight.

Using the Excel equation value the estimated molecular weight is *72 258* Daltons.

Using the 2 decimal place approximation, the molecular weight is *72 444* Daltons.

Thus the hand drawn graph, done with precision, can give a reasonably accurate estimation of molecular weight.

Remember that this is only an estimate. Precise molecular weight determination would require more extensive sequence and chemical analyses. Repeating measurements on different gels and samples would improve accuracy.

PRACTICAL HINTS AND TIPS FOR SDS-PAGE:

Handling of gels and reagents: Gloves should be worn at all stages where reagents and gels are handled, not only for personal protection but also to avoid unnecessary introduction of keratins (skin proteins) into the sample or running buffers, and so on.

Sample preparation:

- Sample buffer usually contains volatile reducing agents such as 2-mercaptoethanol, which will expand on heating.
- To avoid release of pungent vapours into the atmosphere you should place a clamp over the lid of the Eppendorf tube containing the sample during the high temperature incubation in order to prevent the tube lid popping off.

- As an additional precaution, if practical, samples could be boiled and loaded into gels in a fume cupboard (if available).
- It is safer to use a heating block to prepare samples for electrophoresis.

Gel preparation:

- When preparing your own gels make sure that all surfaces coming into contact with acrylamide are clean.
- Plates, combs and spacers should be cleaned with 70% (v/v) ethanol in order to remove traces of grease that might interfere with polymerization.

Sample application:

- Careful pipetting is essential (see Chapter 1).
- When applying the sample into the sample wells, place the tip to the centre of the well avoiding contact with the sides, preferably using gel loading tips which have a thin extension that fits comfortably between the glass plates.
- Do not puncture the stacking gel with the tip as such damage may distort the gel pattern.
- Carefully eject the sample to the second position of the pipette barrel (see Chapter 1), keeping the barrel depressed until the tip has been removed from the sample well altogether; this will avoid suction of the sample back up into the pipette.
- The sample can be seen sinking to the bottom of the well due to the presence of glycerol and bromophenol blue.
- If you cannot see the outline of the sample wells when they are filled with running buffer, draw a point on the glass plate near the centre of each well before assembling the electrode unit or before rinsing out with running buffer.

Vigilance during electrophoresis:

- Make sure that terminals and electrical leads are connected correctly to the apparatus and the power pack. If you plug in the leads the wrong way round your proteins will migrate upwards into the buffer tank instead of migrating downwards into the gel.
- Always check at regular intervals that the buffer level in the inner chamber is above that of the lower glass plate.
- If the levels fall too low there will be no continuity between the electrolyte in the two chambers and the gel and electrophoresis will come to a halt.
- If topping up is required, switch off the power supply before removing the lid.

Gel removal and handling:

- *Switch off* the power before dismantling the apparatus.
- In order to avoid cuts or more serious injury, *never* use a scalpel or scissors to separate the plates from each other after electrophoresis is complete. This could result in damage to the glass plate, airborne chips of glass or scalpel blades.
- Always use either a plastic wedge that fits snugly between the plates or a thin blunt instrument (e.g. a spatula) to carefully separate the plates for gel removal.
- The gel can be lifted by gently picking it up via two corners along its length. Alternatively it can be floated off the glass plate into the fixing or staining solution.

Data analysis:

- As shown in the worked example, always plot $\log_{10}$ of the molecular weights of the standards as an exponential curve produced from a plot of actual molecular weight against Rf, from which the molecular weight would be estimated with less accuracy.
- Distance moved should always be measured between the same reference points. For example, the point of origin should be the start of the stacking gel and the distance moved should be to the middle of the polypeptide band.
- Bear in mind that inaccuracies could result especially from manual measurements using a ruler, for which accuracy would be 0.5 mm at best.
- Repeated measurements on different gels would help to improve accuracy.

Interpretation:

- When asked to tentatively identify a protein based on its size, *do not* assume that a protein having the molecular weight of one of the standards provided is necessarily the same protein as the standard.
- Standards are provided simply to allow the construction of a calibration curve for the estimation of molecular weights.
- SDS-PAGE will only give an approximation of the molecular weight of a polypeptide. Accurate determination would require a more complete sequencing and chemical analysis.
- If a protein has a sub-unit structure, the subunits will be separated from each other. Thus, a protein that exists in its native form as a homodimer (that is, two identical subunits) with a total molecular weight of 100 kDa, which could be determined by size exclusion chromatography (Chapter 7),

would migrate as two polypeptides of approximately 50 kDa each on SDS-PAGE.
- It should also be noted that not all proteins migrate in a log linear fashion. For example, anomalous migration is often observed with glycoproteins due to the increased frictional drag caused by covalently linked oligopeptide chains projecting out from the linearized polypeptide chain.

6.3 Other Electrophoretic Techniques Applied to Proteins

A number of other applications involving the use of gel electrophoresis in protein separation are outlined below. The reader is referred to the reference list for further reading.

6.3.1 Non-Denaturing Gel Electrophoresis

In this approach the protein sample is *not* denatured by boiling in SDS and reducing agent. Instead it is applied directly to the gel (polyacrylamide or agarose) and separated according to size, shape and net charge. An advantage of this technique is that proteins are in their native conformation and retain their ability to form complexes with other proteins. After staining, spots can be eluted and separated under denaturing conditions by SDS-PAGE in order to analyze further the components of any protein complexes. Also, in some applications of non-denaturing gels it is possible to measure enzyme activity in specific polypeptide bands or complexes (e.g. acid phosphatase, dehydrogenase and protease activity). In this case, the electrophoresed gel is incubated with a solution containing substrates for the enzyme of interest and activity is detected by the measureable appearance of a coloured reaction product in the area of the gel to which the enzyme has migrated.

6.3.2 Cellulose Acetate Electrophoresis

A porous matrix of cellulose acetate is commonly used in the separation of blood plasma proteins in clinical diagnostic tests using short separation times of around 40 min. This form of electrophoresis uses a phenomenon known as electro-osmotic flow, caused by the interaction of cations in the buffer with negatively charged groups on the matrix, which can result in the movement of weakly anionic biological macromolecules (e.g. antibodies and other serum proteins) moving towards the cathode.

6.3.3   Western Blotting

In this technique, gels containing proteins separated by either SDS-PAGE or non-denaturing PAGE are transferred electrophoretically from the gel on to a protein binding membrane (e.g. nitrocellulose) which is in contact with it. The resultant western blot is, in effect, a replica of the original gel pattern on the surface of the membrane. However, the membrane bound proteins are now more accessible for probing with antibodies that recognize specific proteins than they would have otherwise been in the gel. This technique has the advantage that it can detect the presence and amount of very low abundance proteins in complex mixtures.

6.3.4   Isoelectric Focusing (IEF) and 2-Dimensional Polyacrylamide Gel Electrophoresis (2D-PAGE)

In IEF samples are prepared in a buffer containing a neutral detergent (e.g. CHAPS), reducing agent (e.g. dithiothreitol), a denaturing agent (e.g. 8 M urea) and pH gradient forming molecules called ampholines. They are then applied to a special type of polyacrylamide gel strip containing an immobilized pH gradient (IPG strips) and separated according to their net charge until all polypeptides reach a point where their net charge is zero, which is known as their isoelectric point (pI; see Chapter 7). While this technique can be useful for the separation of proteins (or protein isoforms) of similar size, but slight differences in net charge, it is much more powerful when used in combination with SDS-PAGE as discussed below. This process is shown schematically in Figure 6.4.

In two dimensional polyacrylamide gel electrophoresis (2D-PAGE), IPG strips containing proteins separated by net charge are then equilibrated with SDS containing denaturing sample buffer and separated by differences in size in a second dimension of SDS-PAGE. This approach, which is shown diagrammatically in Figure 6.4, increases the resolution of complex mixtures compared to either IEF or SDS-PAGE alone, and is a major technique used for analyzing the protein profile (proteome) of cells, tissues and subcellular fractions (e.g. nuclei, mitochondria, cytosol, and so on). Identification would require further analysis on individual polypeptides such as mass spectrometry. Images of a cell extract separated by both SDS-PAGE and 2D-PAGE are shown in Figure 6.4.

6.4   Separation of Nucleic Acids by Gel Electrophoresis

Nucleic acids are negatively charged, due to the phosphate groups in the sugar phosphate backbone. As the net charge is similar, due to the regular repeats of

Figure 6.4 *Principles of two-dimensional polyacrylamide gel electrophoresis (2D-PAGE). Shown are schematic representations of a typical isoelectric focusing (IEF) gel, showing how a few polypeptide bands might focus on an IPG strip at their isoelectric points (a). If this IPG strip were to be equilibrated with SDS-PAGE denaturing sample buffer and applied to the top of a SDS-PAGE (b), it can be seen that a number of polypeptide spots of different sizes could be resolved. An example of a real 2D-PAGE separation of a mouse neuronal cell extract in which the polypeptide profile is revealed by silver staining is shown in (c). An example of the same extract separated by only SDS-PAGE then stained with Coomassie blue is shown in (d). It is clear from these examples that many polypeptides can have a similar size or isoelectric point, but they can be resolved from one another if a combined charge and size separation approach is used (i.e. 2D-PAGE).*

phosphate groups, polynucleotide chains vary mainly in length. Therefore, DNA and RNA molecules will migrate in an electric field towards the anode in an electrophoresis gel according to their size. The principles of migration are similar to those for SDS-PAGE except that agarose gels with larger pore sizes are more commonly used for the separation of large DNA or RNA molecules. Thus, the longer DNA sequences will migrate more slowly than the shorter sequences.

Manipulation of the agarose concentration determines the effective fragment size separation range, lower concentrations (e.g. 0.3% w/v) being better for the resolution of larger DNA size ranges (e.g. 1000–70 000 base pairs) and higher concentrations (e.g. 2% w/v) being more suitable for smaller size ranges (e.g. 100–5000 bp). Resolution of even smaller DNA molecules may require the use of polyacrylamide (typically 29 : 1 acrylamide: *bis*-acrylamide ratio). For example a 20% (w/v) polyacrylamide gel will allow the separation of very small DNA fragments between 5 to 100 base pairs in length, whereas a 3.5% (w/v) polyacrylamide gel would resolve DNA sequences of approximately 100–1000 base pairs. As polyacrylamide gel preparation has been covered already in this chapter, we will now focus on the preparation and use of agarose gels.

Note: DNA electrophoresis may use toxic chemicals, particularly during the gel staining procedures (e.g. ethidium bromide). You must therefore ensure that you prepare, handle and dispose of toxic reagents in compliance with local health and safety policy. However, alternative dyes such as RedSafe™ have been developed recently which are non-carcinogenic but may cause skin and eye irritation. It has the advantage that it can be visualized not only under UV light but also with visible light, reducing the risk of UV damage when the DNA is to be manipulated in further experiments. It can detect as little as 50 ng of double stranded DNA in a gel band, which is comparable to the sensitivity of ethidium bromide.

6.4.1 Sample Preparation

DNA or RNA is prepared by specific extraction procedures (see Section 5.2.2). Before preparing your sample for electrophoresis, it is important to estimate the nucleic acid concentration so that precise amounts of nucleic acid can be loaded onto the gel. This is typically done by estimating the ratio of absorbance 260–280 nm, measured using a spectrophotometer (see Section 3.8.1).

6.4.2 Preparation and Running of Agarose Gels

Agarose gels tend to be run in a flat bed electrophoresis system in which the gel, which is completely submerged in electrolyte (Table 6.2), is run horizontally (as

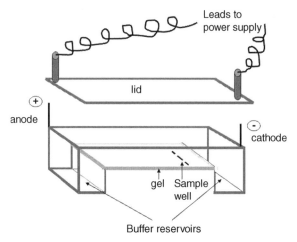

Figure 6.5 Schematic diagram of a typical apparatus for agarose gel electrophoresis. A typical apparatus comprises two buffer reservoirs either side of a raised platform onto which the agarose gel is polymerized. The samples are applied into the sample wells shown and electrophoresed in a horizontal direction. Nucleic acids migrate towards the anode due to their intrinsic negative charge due to the phosphate groups. The buffer tank is filled until the gel is submerged in the electrolyte.

opposed to the vertical direction used for the majority of protein separations). A typical gel apparatus is shown schematically in Figure 6.5. Agarose gels tend to be much thicker than polyacrylamide gels as thinner strips of agarose would be very fragile. The agarose is mixed with appropriate buffer components and dissolved by heating, then allowed to form a gel on cooling. As for protein gel electrophoresis, sample application wells are formed by the insertion of a sample well comb before the gel is able to set. This is removed carefully and the wells filled with electrolyte prior to sample application. As for SDS-PAGE, the sample is loaded through the electrolyte aided by the presence of glycerol in the sample buffer (Table 6.2) which makes the sample sink to the bottom of the well. The current is applied at the recommended level (this varies from one apparatus to another) and the gel allowed to run until the tracker dye (e.g. bromophenol blue) has moved at least 3/4 of the gel length.

6.4.3 Staining of Gels to Detect Nucleic Acids

Gels are carefully removed from the gel tank with gloved hands and immersed in a solution containing the fluorescent nucleic acid binding dye ethidium bromide or a suitable alternative, which intercalates between paired nucleotide of sections of double helix in DNA, for approximately 5 min. The gel is then carefully transferred to a UV trans-illuminator which may be part of an imaging instrument which encloses it in a sealed unit. It is important to wear protective glasses, particularly if a sealed imager is not available, as there is a risk that your eyes may be over exposed to UV. Bands can be observed by a special camera that detects UV light and a digital or photographic image recorded. DNA or RNA molecules are visible as fluorescent bands that migrate according to size differences. Running gels with a series of standards (polynucleotides of a size range that covers the range of resolution of a particular gel) in parallel allows the estimation of molecular weight from a standard curve similar to that used in SDS-PAGE. However, remember that in the case of nucleic acids, molecular weight is measured in base pairs (bp). A more precise evaluation of molecular weight would require sequence analysis to take into account the exact base composition.

6.5 Applications of Gel Electrophoresis of Nucleic Acids

6.5.1 DNA Fingerprinting and PCR

This involves the electrophoretic separation of DNA fragments produced either by reverse transcriptase polymerase chain reaction (RT-PCR) or restriction endonuclease mediated digestion of DNA. In the case of RT-PCR the amplified product (amplicon) is often eluted from the gel and sequenced to confirm that the appropriate gene sequence has been amplified. When the digested DNA is from whole DNA it can be used to determine the DNA fingerprint of an individual or species or to identify polymorphisms in genes (i.e. changes in the sequence of DNA that cause changes in endonuclease cleavage resulting in a different DNA fragmentation pattern). An image of a typical stained agarose gel is shown in Figure 6.6.

6.5.2 Analysis of RNA

Total cellular RNA, ribosomal RNA and purified messenger RNA molecules can be separated using agarose gel electrophoresis. The quality of total RNA preparations can be assessed by the appearance of the larger stronger staining bands corresponding to ribosomal RNA subunits (Figure 6.6). However, it is

Figure 6.6 *Examples of agarose electrophoresis gels stained with ethidium bromide. The agarose gel on the left hand side shows the molecular weight standards (Std: a ladder of polynucleotides of different sizes) and the PCR amplification products of two different mouse gene sequences of similar length (seq1 and seq2). Both sequences have a size of approximately 500 bp. The gel on the right shows the separation of total RNA from three different plant cell extracts. A consistent pattern can be seen in each sample, the two strongest bands representing the ribosomal RNA sub units. The streaky 'tail' below the smaller (lower) of the two bands reflects the many different sizes of messenger RNA sequences that would be expected to be present in a cell or tissue extract. The two gels are not directly comparable as a much lower agarose concentration was used for the gel on the right.*

important to note that under conditions used for double stranded DNA separation, RNA molecules may exhibit secondary structure that causes abnormalities in their electrophoretic migration, making accurate molecular weight determination more difficult. Secondary structure can be eliminated by performing denaturing gel electrophoresis for which the samples are pre treated by heating in the presence of formamide or glyoxal and electrophoresed using buffers containing formaldehyde.

6.5.3 Pulsed Field Gel Electrophoresis (PFGE)

This technique allows separation of DNA sequences of up to 12 Mb in length and allows researchers to gain structural information about very large regions of genomic DNA (i.e. DNA containing both coding and non-coding sequences). These regions of DNA are produced by digesting genomic DNA with restriction endonucleases that cut at very rare cleavage sites containing eight base pairs, resulting in the production of much larger fragments than would be produced by more commonly used restriction enzymes, which cleave at sites between four to six base pairs in length. PFGE is different from the more conventional method described earlier in that it uses at least two alternating electric fields. These fields are arranged at different angles in such a way that results in very effective size based separations, due to the fact that larger DNA molecules take longer to reorientate to the field direction change than do smaller ones.

6.5.4 Identification of Specific Sequences Using Hybridization Techniques

After conventional agarose gel electrophoresis, separated nucleic acids can be transferred onto a membrane filter (usually nylon or nitrocellulose) to produce a replica of the gel separation pattern. To facilitate transfer to the membrane, the gel is first of all soaked in HCl (which randomly cleaves DNA into smaller pieces) then with alkali (which denatures the double stranded DNA to single stranded polynucleotide chains) and then neutralized. Alternatively, DNA can be fragmented at specific points along its polynucleotide sequence using enzymes termed restriction endonucleases, and the resultant fragments separated by agarose gel electrophoresis.

Notes on Restriction Endonucleases

- These are enzymes that can cleave double stranded DNA along both strands. 'Restriction enzymes' recognize specific sequences at which they are able to cleave the DNA without damaging the bases.

- This approach has been a cornerstone in recombinant gene technology, as the resultant DNA pieces can be joined up (ligated) with other DNA segments cut with the same restriction enzyme, using DNA ligases.

- Examples of restriction enzymes and their cleavage sites on double stranded DNA are shown below.

- The shaded and unshaded areas represent the ends of the two cleavage products.

- Note that the ends can be uneven ('sticky ends') or straight ('blunt ends').

- As the occurrence of a specific cleavage site is unique to each gene, this approach is also applied in DNA fingerprinting in forensic science and in the analysis of restriction fragment polymorphisms which can help to identify gene mutations.

The gel is then placed in close contact with the membrane filter and several layers of filter papers which soak up the buffer and nucleic acids in the gel through the membrane by capillary action, leaving the nucleic acid sequences bound to the membrane as they come into contact with it. This process can be speeded up by using positive pressure, a vacuum or an electric field.

The blot is then fixed by heating in a drying oven before specific stretches of DNA or RNA are identified by incubation of the blot with complementary nucleotide sequences with which they will hybridize (bind). These probes are either radiolabelled during synthesis (detected by autoradiographic exposure onto X-ray film) or have a non-radioactive label (e.g. digoxigenin) that can be detected by a chemical reaction.

This kind of approach can be used for probing DNA (Southern blotting) or RNA (Northern blotting). The combined use of gel electrophoresis and blotting gives information about the size of a specific sequence. However, if only detection is required, an alternative to the above is to use dot or slot blotting, in which the nucleic acid extracts are pipetted directly to the membrane through round (dot) or slot-shaped template holes, then probed as for the Northern and Southern blots.

6.5.5   DNA Sequencing

The sequence of DNA can be determined by performing a procedure known as the Sanger reaction on a given gene or part of a gene sequence. This involves the use of single stranded DNA as a template for DNA synthesis in the presence of low levels of radiolabelled or fluorescently labelled dideoxynucleotides which, as they are unable to form the phosphodiester bonds in the DNA backbone, act as a chain terminator. If separate reactions are performed with all four dideox-ynucleotides (ddA, ddT, ddG and ddC) at low concentrations, some polynucleo-tide chains will lengthen as normal, whereas others will terminate at each stage that the normal deoxynucleotide is present, leading to a series of chain lengths for each dideoxynucleotide corresponding to each position where the corresponding deoxynucleotide is found. Such chains are separated on denaturing polyacrylamide gels and the bands detected by autoradiography (radioisotopes) or by laser scanning (fluorescence). The sequence of the template is then deduced from the position of the bands (which represent chains terminated at each point that the corresponding deoxynucleotide should be present) in each of the four sample wells from top (3') to bottom (5') ends of the gel. Remember that the sequence determined initially is the complementary sequence to the original template, thus the true sequence is determined by its complementary base pairing (e.g. A-T, G-C). A schematic representation of a sequencing gel is shown in Figure 6.7,

Figure 6.7 *Schematic representation of a DNA sequencing gel. Shown is a schematic representation of a typical gel pattern detected by autoradiography of a sequencing gel. Above each sample well is an indication of which dideoxy nucleotide was added in the reaction described in Section 6.5.5. Each band represents a chain length at which the polynucleotide synthesis came to a halt due to the addition of a dideoxynucleotide instead of the deoxynucleotide that would be added during normal DNA synthesis. The positions of the bands in the four different banding patterns can then be used to determine the position of each consecutive nucleotide in the polynucleotide sequence. The sequence of the original template is then derived from the known pattern of base pairing, as shown.*

together with an indication of how the sequence is determined from the band profile on the gel.

6.6 Summary

- Gel electrophoresis is a means of separating proteins and nucleic acids in an electric field through a three-dimensional gel matrix.
- Separation is dependent on differences in mass, shape and charge of the separated molecules and on the resistance experienced during electrophoretic migration.
- Polyacrylamide gels are used mainly for protein separations but also for small polynucleotides, whereas agarose gels are the norm for larger nucleic acid sequences.
- Electrophoresis has major applications in the characterization of isolated proteins and nucleic acids (e.g. determination of molecular weights and purity of purified proteins, nucleic acids, PCR products and in DNA sequencing).
- It can be combined with blotting techniques that allow probing of protein or nucleic acid samples transferred onto membrane filters for specific components, using antibody or polynucleotide probes.

7

CHROMATOGRAPY

7.1 General Introduction

Chromatography was first developed by the Russian botanist Mikhail Tswett in
the early 1900s when he produced a colourful separation of plant pigments. The
term chromatography translated literally from the Greek words *chroma* and
graphein means to 'write with colours'. The technique remained peripheral to
mainstream analytical techniques until the 1930s, when paper and thin layer
chromatography (TLC) became popular. In the following decades many novel
chromatographic techniques were developed, making chromatography the most
widely used analytical technique in the biosciences.

The diversity of chromatographic methods means that chromatography can be
used to purify any soluble (or volatile) compound if the correct adsorbent
material, mobile phase and operating methods are employed. It can separate
complex mixtures with great resolution; components such as proteins that may
vary by only a single amino acid can be separated by chromatography. These
components are also immediately available for identification and quantification.
If an important component has been identified, the whole chromatographic
procedure can be 'scaled up' from a small scale analytical run to a preparative
run. Also, the conditions employed for chromatography are typically not severe
(gas liquid chromatography is an exception), which allows for the analysis of
sensitive components.

7.2 The Theory of Chromatography

All compounds have different structures and the properties of these different
structures will influence the interaction a compound will have with its immediate
environment. If a compound is placed into an environment containing two
immiscible phases, the properties of the compound's structure will determine
which of the phases (or both phases) the compound will preferentially interact
with. A distribution equilibrium (K_D; also called the partition coefficient) will be
established which is a measure of a compound's capability to interact with the

Basic Bioscience Laboratory Techniques: A Pocket Guide, First Edition.
Philip L.R. Bonner and Alan J. Hargreaves.
© 2011 John Wiley & Sons, Ltd. Published 2011 by John Wiley & Sons, Ltd.

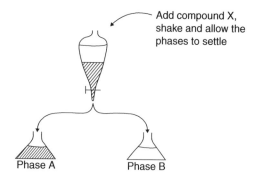

Add compound X,
shake and allow the
phases to settle

Phase A   Phase B

The amount of compound X in each phase can be
measured and the partition coefficient (K_D) determined

$$(K_D) = \frac{\text{the concentration of compound X in phase A}}{\text{the concentration of compound X in phase B}}$$

Figure 7.1 *Separating funnels can be used to determine a partition coefficient (K_D).*

two immiscible phases. It is a ratio of the concentration of a compound in one phase divided by its concentration in the other phase.

Compounds can be separated from each other in a simple laboratory apparatus called a separating funnel, into which two immiscible liquid phases have been placed. The separating funnel is shaken to evenly distribute the compound between the two phases and allow equilibrium to be established. When the phases have separated they can be isolated into separate collecting vessels. The concentration of the compound in each of the phases can be analyzed (see Figure 7.1) and the partition coefficient (K_D) determined.

Separating funnels produce one partition event, which may be insufficient to gather the majority of the compound of interest into one phase. To improve the separation and the yield of the compound of interest, the experiment can be repeated with fresh liquid. Using separating funnels this is a time consuming process but chromatography columns can reproduce these events many times in quick succession. In chromatography the two immiscible phases are the *mobile phase* (solvent) which carries the sample over and through the *stationary* phase.[1] These phases may be combinations of liquid and liquid, gas and liquid or liquid and solids. Mixtures of compounds introduced into this flowing system partition between the two immiscible phases and they will be separated if they have different partition coefficients. Sample components that preferentially partition onto the stationary phase will spend a greater amount of time on the column. These components are separated from other components that stay predominantly in the mobile phase as these components will pass through the column at a quicker rate. The detector at the end of the column will respond proportionally to the concentration of the compound in solution, marking the presence of the compound as a peak with a characteristic retention time (or volume). Every compound applied to a chromatography column will emerge as a 'normal' (Gaussian) distribution (see Figure 7.3 and section 4.3.1).

Figure 7.2 *Theoretical plates in chromatography columns.*

7.2.1 The Partition Coefficient (KD) in Chromatography

$$K_D = \frac{\text{the concentration of a compound in the mobile phase}}{\text{the concentration of a compound in the stationary phase}}$$

The length of column where an individual equilibrium between the sample in the mobile phase and the stationary phase takes place is called a theoretical plate (there will be a total of N theoretical plates in a column; see Figure 7.2 and Worked Example 7.1). The length of the column containing a theoretical plate is called the plate height (H). The more times a compound can establish equilibrium on the column, the more likely it will be resolved from other components. Therefore a good chromatography column will have a large number of theoretical plates (see Worked Example 7.1).

Although theoretical plates are a useful numerical measure of a column's efficiency it is important to remember that they do not actually exist. The plate model of chromatographic separation assumes an almost instant interaction between a compound in the mobile phase and the stationary phase. The plate model also assumes that the movement of compounds down the column is by the transfer from one theoretical plate to the next.

However, compounds applied to a column take time to interact with and percolate through the stationary phase. The **rate theory** of chromatography takes into account other factors that contribute to the final elution profile of a compound in chromatography.

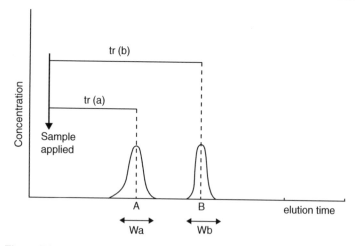

Figure 7.3 *The elution profiles of compounds A and B on a chromatography column. The time the compounds have been retained on the column (tr) and the peak width (W) can be measured.*

- The width of a chromatographic peak (see Figure 7.3) is linked to the flow rate of the mobile phase (u).
- The molecules of any compound applied to a column will take different random paths through the stationary phase which contributes to peak broadening (eddy diffusion) (A).
- The concentration of any compound is greatest at the centre of the peak and this concentrated material will tend to diffuse out toward the edges, which will also contribute to peak broadening (longitudinal diffusion) (B).
- When a compound has a strong interaction with the stationary phase and the flow rate is too high, any compound remaining in the mobile phase will be swept ahead of the compound interacting with the stationary phase, resulting in peak broadening (resistance to mass transfer) (C).

The van Deemter equation relates these factors to the height equivalent to a theoretical plate (HETP) of a column and stationary phase.

$$\text{HETP} = A + \frac{B}{u} + Cu$$

Worked Example 7.1: Resolution, Peak Width and Theoretical Plate Number

Two compounds were applied to a 15 cm column. The retention time for compound A was 5 minutes and the base line peak width was 12 seconds. The retention time for compound B was 8 minutes with a peak width of 10 seconds.

Chromatographic columns are judged by their ability to provide baseline separations of different components in sharp peaks (see Figure 7.3). Two important parameters are resolution (R) and the peak width (W).

$$R = \frac{[2(tr(b) - tr(a))]}{[W(a) + W(b)]}$$

$$R = \frac{[2(480 - 300)]}{12 + 10}$$

$$= 16.36 \text{ sec}$$

- A number greater than 1.5 usually means baseline separation. Clearly these peaks are well resolved and a visual check of a chromatogram is always recommended.
- The efficiency of the separation by a chromatography column of individual components is measured in theoretical plate number (N).

$$N = 16(tr/W)^2$$

For compound A

$$N = 16(300/12)^2$$
$$= 10\,000$$

For compound B

$$N = 16(480/10)^2$$
$$= 36\,864$$

Chromatography columns act as if they have different plate numbers for different compounds.

- When two peaks elute closer together the peak width at half peak height ($W_{0.5h}$) can be used to calculate the theoretical plate number.

$$N = 5.54(tr/W_{0.5h})^2$$

The height equivalent to a theoretical plate (HETP) allows comparisons to be made between columns of different lengths (L) or the same column run under different conditions.

$$HETP = L/N$$

Column efficiency should improve with an increase in N or a decrease in the HETP.

7.3   Factors to Consider in Chromatography

- All compounds elute from a chromatographic column as a 'normal' (Gaussian) distribution and they will have a characteristic retention factor (time, volume or distance moved) (see Figure 7.3).
- Identification of compounds in a sample can be achieved by comparing the retention time of a known standard to the retention time of compounds in the sample (this does assume some prior knowledge of the type of compounds to be separated).
- Quantification can be achieved by chromatographic separation of increasing amounts of a standard compound and measuring the peak area of the eluted peak.
- A calibration graph can be constructed of peak area (y-axis) against concentration (x-axis) which can be used to quantify the compound in an extract.
- Most modern chromatographic systems have data management devices (computers or integrators) linked to the detector (see Figure 7.7), which can be used to measure the peak areas and produce calibration graphs.
- However, if the system is linked to a chart recorder the peaks can be cut from the paper with scissors and weighed to construct a calibration graph.
- Applying crude samples to a column will not only complicate the analysis but can lead to some components permanently bonding to the resin, reducing the column's capacity (the amount of sample which can be applied to the column without altering the resolution). Solid phase extraction (SPE) is a popular procedure to remove 'column damaging' compounds and to enrich the components of interest. Cartridges of various dimensions can be purchased and used with a syringe to quickly clean up the sample mixture. If a compound is known to elute from a reversed phase C_{18} (ODS: octadecylsilane) analytical column at 50% (v/v)^2 methanol, a C_{18} SPE cartridge can be used in the sample preparation (see Worked Example 7.2 and section 7.5.1).
- To overcome any variation in sample preparation and to counter the small changes in the amount of sample applied to the column, a known amount of a compound can be added to the extract. This internal standard should be chemically related to the compounds of interest, but it should also ideally elute in a different part of the chromatogram. Variation in the extraction or injection process can then be compensated for by reference to the constant area of the internal standard.
- Use the highest grade of solvents and water available to make up chromatographic reagents.
- Always filter the reagents through a 0.2 μm filter to remove particulate matter (and bacteria) which could clog up the pores in the column.
- To protect the column from adverse pressure surges, set a maximum pressure limit (below the maximum allowed) on the chromatographic system.

Worked Example 7.2: A typical procedure for the preparation and use of a C18 SPE column in sample preparation

- The SPE cartridge is first washed in methanol before being equilibrated in 10% (v/v) methanol.
- Methanol is added to the sample to a final concentration of 10% (v/v) before application to the SPE cartridge.
- The SPE cartridge is then washed with 10% (v/v) methanol to elute components that do not bind to the resin.
- The sample can then be eluted in 70% (v/v) methanol and dried or diluted before being applied to the analytical C_{18} column.
- Any damaging components will remain bound to the SPE cartridge rather than binding irreversibly to the expensive analytical C_{18} column.
- The SPE cartridge can then be disposed of in an appropriate fashion.
- A syringe is fine for individual samples but for processing large numbers of samples the SPE cartridges can be loaded into proprietary SPE racks, which use vacuum pumps to quickly load and elute the samples.

7.4 Methods Used to Elute Samples in Chromatography

- **Isocratic elution:** the mobile phase remains constant for the elution period (e.g. 2–5 column volumes). At the end of the isocratic wash period, the mobile phase may be altered for another isocratic wash (see Figure 7.4a). This is a useful technique to employ when components elute close together in a gradient elution.
- **Gradient elution:** the mobile phase is gradually altered to increase or decrease one of its components (see Figure 7.4b).
- **Displacement elution:** involves washing the column with a molecule that has a higher affinity for the ligand on the stationary phase than the molecule bound to the resin. Displacement elution can be used in ion exchange chromatography (see Section 7.8) and affinity chromatography (see Section 7.10). The eluted component migrates down the column as a front rather than the normal distribution seen in gradient or isocratic elution.

7.5 Different Types of Chromatography and What Properties Can be Used to Separate Molecules

Chromatography is a method that relies on the *differences* in partitioning behaviour (K_D) between a flowing mobile phase and a stationary phase to separate the components in a mixture (see figure 7.5). Molecules can be

Figure 7.4 *Two common methods used to elute samples from chromatography columns.*

Figure 7.5 *Different types of chromatography.*

separated from one another if there are differences in one or more of the following properties;

- molecular size and shape
- surface charge (the presence of positive and negative charges)
- polarity (a measure of how hydrophobic or hydrophilic a compound is)
- volatility (how easy does it become a gas).

The differences in these properties can be exploited by chromatographic methods to separate components in a complex mixture (as outlined below).

7.5.1 Adsorption Chromatography

Adsorption chromatography (see Section 7.6) allows for sample compounds to interact with adsorption sites on the stationary phase. In adsorption, the sample interacts only with the surface of the stationary phase (mainly by van der Waal interactions) and is not absorbed, which includes entering into the structure of the stationary phase. The interaction at the surface allows an equilibrium to be established with the mobile phase. This can be a solid/liquid interaction for example, thin layer chromatography (TLC) or a solid/gas interaction for example, gas/liquid chromatography (GLC).

7.5.2 Partition Chromatography

Partition chromatography (see Section 7.7) allows for a sample compound to dissolve (partition) and establish an equilibrium between the liquid bonded to the stationary phase and the mobile phase. This is usually a liquid/liquid interaction for example, normal and reversed phase chromatography.

7.5.3 Ion Exchange Chromatography

Ion exchange chromatography (see Section 7.9) allows for samples with a positive or negative charge to interact with an opposite charge bonded to the stationary phase and establish an equilibrium with the mobile phase, for example ion exchange chromatography of proteins.

7.5.4 Size Exclusion Chromatography

Size exclusion chromatography (see Section 7.10) allows for samples with unique size and shape to interact with the porous spaces contained within the stationary phase and establish an equilibrium with the mobile phase, for example size exclusion chromatography of proteins.

7.5.5   Affinity Chromatography

Affinity chromatography (see Section 7.11) allows for a sample with a unique structural feature to interact with a complementary structure bonded to the stationary phase and establish an equilibrium with the mobile phase, for example affinity chromatography of proteins.

$$(R_f) = \frac{\text{the distance moved by the sample}}{\text{the distance moved by the solvent front}} = \frac{x}{y}$$

Figure 7.6 Thin layer chromatography.

7.6   Thin Layer Chromatography (TLC)

Thin layer chromatography (TLC) is a solid-liquid chromatographic technique. The stationary phase is a micro granular solid that sample components can adsorb to, thus establishing equilibrium with the liquid mobile phase that passes over the stationary phase. The most commonly used solids in this form of chromatography are silica gel (SiO_2), alumina (Al_2O_3) and cellulose. Polar groups on the surface of these solids form weak interactions (H-bonds and dipole interactions) with samples dissolved in an organic liquid. Good solids for adsorption chromatography should be insoluble in the mobile phase, inert to reaction with the sample and have a small particle size to increase the available surface area.

Thin layer chromatography is a sensitive, fast, simple and relatively inexpensive analytical technique. Several samples can be run on the same plate with standards to help in the identification of unknown compounds. It is usually conducted in a planar format with the micro granular solids (adsorbent) bonded to a flat plastic or glass plate. This technique can also be scaled up into a column format to separate larger amounts of sample.

Worked Example 7.3: Thin layer chromatography (TLC) of amino acids

In a planar format, TLC involves spotting the sample onto the adsorbent near one end of the TLC plate. The plate is placed into a tank containing a small volume of solvent. The tank is covered and the solvent moves through the stationary adsorbent by capillary action (see Figure 7.6).

- With a pencil lightly score a sample application line (origin) 1 cm from the bottom of the cellulose TLC plate and mark application points on this line.
- Using a capillary applicator spot 10 μl of the amino acid sample onto an origin point. The best results are achieved by taking time to apply a small concentrated spot rather than a large dilute spot. Hold the applicator in one hand and a hot air blower (hair dryer) in the other, then dot and dry, dot and dry until all the sample has been applied.
- Place the plate in the TLC chamber. The solvent (butanol: acetic acid: water 46:10:26 mixed 50:20 with acetone) level in the chamber should be just below the point of sample application on the TLC plate.
- The solvent will move through the plate by capillary action.
- When the solvent front reaches 1 cm from the top of the plate, remove the plate from the tank and mark the solvent front.
- Dry the plate with a hot air blower and stain the plate with 0.2% (w/v) ninhydrin in acetone (take care to wear gloves).
- Heat the plate with the hot air blower (the ninhydrin reaction with amino acids works best at 70 °C) and record your observations by (a) noting the colour (different amino acids produce different shades of purple and proline produces a yellow colour) and (b) determine the R_f value.

R_f = Retention factor (relative to the solvent front).

$$R_f = \frac{\text{the distance moved by the sample (centre of the spot)}}{\text{The distance moved by the solvent front}} = \frac{X}{Y}$$

Note:

R_f is a number between 0 and 1.0. It has no units, so the distances can be measured in inches, cm or mm.

Rf values will vary so always quote the stationary phase, solvents and temperature used for the experiment.

Compounds with similar Rf values will migrate to the same space on the TLC plate. Therefore some spots on the plate may contain more than one component.

The spots on a TLC plate can be scraped from the plate and the components recovered in a solvent. Another chromatography separation can then be conducted to separate multiple components.

7.7 High Pressure Liquid Chromatography (HPLC)

As mentioned above, the same adsorbent and solvents used in TLC can be used in a column format to separate larger amounts of sample. This is usually done by filling a glass column with a stopper at the bottom with adsorbent and after applying the sample the column is eluted with solvent and fractions are collected until the sample elutes.

The particle size of the adsorbent in these columns is relatively large (50 μm) which means there is adequate space between particles for the solvent to percolate through the column under the effects of gravity (hydrostatic pressure). Using these large particles, adequate resolution of complex mixtures is achieved using small volumes of concentrated samples applied to columns with large volumes of

Figure 7.7 *Schematic diagram of an HPLC system.*

158

Key to figure 7.7

A. Solvent reservoirs are usually made from glass. There is usually a filter at the end of the solvent inlet to help prevent particles from entering the chromatography system.

B. High pressure pumps are required to deliver the mobile phase (at flow rates up to 10.0 ml min^{-1}) through the stationary phase at a pressure of at least 6000 psi.

C. The sample is normally loaded onto the column using a fixed volume injection loop. Manual or automatic injection systems can be included in a system set up.

D. HPLC columns come in a variety of different formats. They are traditionally made from steel tubing with an internal diameter (i.d.) of about 4.0 mm, approximately 5–50 cm in length and are normally run at flow rates up to 5 ml min^{-1}. Microbore columns (i.d. 2.0 mm) operate at flow rates up to 0.2 ml min^{-1} and capillary columns 5.0 cm in length (i.d. 0.2 mm) have nl min^{-1} flow rates. The reduced flow rates from the smaller diameter columns are ideal when interfacing with a mass spectrometer detector.

Silica based resins are unstable in alkali environments (>pH 8.0). Chemically modified silica resins have been produced to overcome this problem and allow separations to take place at elevated pH values. Microparticulate resins (3–10 μm dia) are the most popular form of silica stationary phase used in HPLC. These resins can be porous, the smaller the pore size the greater the overall surface area of the stationary phase. Other chemical modification of silica has resulted in the production of small silica particles (1–2 μm dia) which are more pH stable and able to withstand elevated pressures. Pellicular resins (30 μm dia) have a solid core to the particle to which is bonded a thin layer of silica (1–2 mm). This porous outer layer allows rapid interactions to take place between the mobile and stationary phases and is popular in ion exchange resins (see Section 7.9).

E. Online spectrophotometers capable of measuring UV/visible (200–800 nm) light are popular detectors. Other online detectors include fluorimeters, radiochemical, electrochemical and mass spectrometers. The detectors are usually linked to a computer or integrator to facilitate the analysis.

F. Fraction collectors can be included to collect valuable components after they have passed through the detection system.

adsorbents. To decrease the column size while retaining the theoretical plate number and the resolving power (see Worked Example 7.1) the particle size has to be decreased. As the particle size decreases (5–2 μm) there is less space for the mobile phase to easily move through. The solvent will no longer percolate through the column due to gravity and requires pressure to force it through the column. In high pressure liquid chromatography (HPLC), mechanical pumps provide the means to force the mobile phase through the stationary phase.

7.7.1 Normal (NP), Reversed Phase (RP) and Hydrophobic Interaction (HIC) Chromatography

Two common chromatographic separations used with a HPLC system are normal phase and reversed phase (RP) chromatography. Small particles of silica (diameter 5–2 μm) provide the solid matrix base for the stationary phase, to which is bonded a monolayer of a polar material (e.g. alkyl amine) in *normal* phase chromatography or a monolayer of organic material for example, (C_{18} aliphatic chain) in *reversed* phase chromatography.

In normal phase chromatography, the polar elements of sample molecules will tend to adsorb to polar material surrounding the solid chemistry of the stationary phase when the mobile phase is non polar for example, hexane. Different compounds will have different levels of polarity with the most polar being retained on the column longer, increasing the retention time. Normal phase chromatography is most useful when it is used with compounds that are predominately non polar but contain an element of polarity for example, phospholipids, steroids and fat soluble vitamins such as vitamin E. Elution can be achieved by increasing the polarity of the solvent but this can also generate problems, as water molecules can then interact with and deactivate the stationary phase. This results in changes in the retention times of the compounds applied, making the reproducibility of the technique very difficult.[3] The changes in retention time can be accommodated by constantly comparing the retention time of the sample to the retention time of an internal standard (see Section 7.3).

Reversed phase (RP) chromatography is the most widespread form of chromatography because of its versatility. A variety of organic compounds (C_5, C_8 C_{18} and phenyl) can be bonded to the stationary phase to facilitate the separation of many different compounds. In RP the initial mobile phase is polar, containing a small percentage (5–10%) of a solvent such as acetonitrile or methanol. When a sample is applied to a RP column there is a liquid/liquid partition between the polar mobile phase and the non polar oily liquid monolayer surrounding the silica bead. Elution is achieved by increasing the proportion of solvent in the mobile phase (e.g. 5–80%). The polar compounds elute earlier than the non polar compounds in the sample. The increase in the proportion of organic solvent in the mobile phase becomes increasingly attractive to any non polar components in the sample. If an unknown mixture of compounds is to be analyzed, the first choice of resin is usually a C_{18} (ODS; octadecylsilane) phase. This phase has proven to be extremely versatile and reproducible. If an adequate separation cannot be achieved with a C_{18} resin, there are many different manufacturers of RP resins with subtly different chemistries which can be subsequently tested for their ability to facilitate the desired separation.

Hydrophobic interaction chromatography (HIC): Proteins contain amino acids with hydrophobic residues (e.g. leucine, tyrosine and valine) which are normally kept away from water molecules. To minimize their contact with water, these hydrophobic amino acids can be buried in the core of a protein's structure or, if they are near to the surface, they can be covered by other amino acid residues. In the presence of high salt concentrations, the surface hydrophobic amino acid residues in a protein's structure can become exposed. These are then able to interact with hydrophobic molecules (e.g. C_8, C_5 or phenyl) bonded to the stationary phase. The strength of binding depends upon how many and what types of hydrophobic amino acids are involved; removing the bound proteins involves decreasing the concentration of salt in the mobile phase.

Hydrophobic interaction chromatography is similar to reversed phase (RP) chromatography, in that organic ligands bonded to a stationary phase interact with non polar hydrophobic groups on the structure of sample molecules. Proteins will bond to RP stationary phases but the high ligand density on RP resins means that the interaction is strong, requiring organic solvents to remove the protein. The proteins will elute but they will no longer easily refold into their normal biologically active structure. The lower ligand density on HIC stationary phases means that proteins can bind and elute while retaining their biological function.

7.8 Gas Liquid Chromatography (GLC)

In a similar manner to reversed phase chromatography, gas liquid chromatography (GLC) involves an organic compound bonded to the stationary phase, effectively forming a monolayer of organic liquid. The stationary phase is

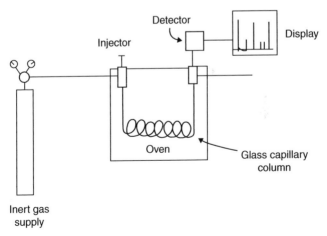

Figure 7.8 Schematic representation of GLC.

usually packed into a glass capillary column (3 mm i.d.), which is then wound onto a reel and inserted into an oven (see figure 7.8). The mobile phase is an inert gas (typically nitrogen) pumped through the stationary phase under pressure.

The volatile sample is injected onto the head of the chromatographic column where it can partition into the liquid bonded to the stationary phase. The sample is transported through the column by the flow of inert, gaseous mobile phase. The oven housing the stationary phase can be programmed to provide a gradient of heat, gradually increasing the temperature (up to 200 °C) over a period of time. Different compounds volatilize (enter the gas phase) at different temperatures, where they are carried by the mobile gas phase to the end of the column and the detector.

Figure 7.9 *The structure of alanine and the change in overall charge with pH.*

7.9 Ion Exchange Chromatography (IEX)

Ion exchange chromatography (IEX) can be used to separate molecules based on their overall charge because it involves an equilibrium between a stationary charged resin and oppositely charged molecules in the mobile phase (usually involving a buffer to maintain the pH). Ion exchange resins can be divided into positively charged anionic resins which attract anions (negatively charged molecules) and negatively charged cationic resins which attract cations (positively charged molecules). The basis of the separation in ion exchange chromatography is that different molecules at any pH value will have different levels of charge.

Strong acids readily dissociate in solution at all pH values whereas the dissociation of a weak acid is dependent on the pH of the solution. Many biological molecules (e.g. amino acids, proteins, nucleic acids and charged metabolites) have weak acid groups in their structures and the charge on the molecule is dependent on the pH (see section 1.6.2). For example, amino acids have both positive and negative weak acid groups in their structures (zwitterions). At a low pH alanine (see Figure 7.9) has an overall positive charge and at a high pH the overall charge is negative. At pH 6.0 the positive and negative charges on alanine are balanced, making the overall charge on the molecule zero. This is called the isoelectric point (pI); at this point the molecule will not bind to an oppositely charged particle (ion exchange chromatography) or it will not move under the influence of an electrical field (see section 6.3.4). Between pH 3–10 alanine has a mixture of positive and negative charge.

Figure 7.10 *Binding and elution from an anion exchange resin.*

Proteins are polymers of amino acids, some of which have weak acid functional groups (aspartic acid and glutamic acid are negatively charged at pH 7.0; lysine and arginine are positively charged at pH 7.0). This means that proteins, like amino acids, have zwitterionic properties. These weak acid charges contribute to a protein's structure, functionality and pI. At 1.0 pH unit below the pI of a protein, it will have sufficient positive charge to bind to a cation exchange resin and at 1.0 pH unit above its pI it will bind to an anion exchange resin. The overall level of binding depends on the net charge at that pH. Molecules can be eluted from ion exchange resins by (a) increasing the concentration of salt (e.g. NaCl) in the mobile phase (see figure 7.10) or (b) raising the pH to the pI of the molecule of interest (cation exchange chromatography) or lowering the pH to the pI of the molecule of interest (anion exchange chromatography) or (c) a combination of (a) and (b).

Silica (porous or pellicular: see the key for Figure 7.6) and polystyrene are used as the solid support in ion exchange resins able to withstand high pressures. Cellulose, dextrans and cross linked acrylamide are popular supports for low pressure

Blue Dextran

Blue Dextran is excluded
from the gel. It will elute
in the void volume (Vo)

Bovine serum albumin

Bovine serum
albumin can
occupy the space
around the beads
and some of the
space within the
beads

Vit B$_{12}$

Vitamin B$_{12}$ is
totally included in
all the space of the
resin. It will elute in
the total volume (V$_t$)

Figure 7.11 *The paths taken by three components in size exclusion chromatography (Blue Dextran ■; bovine serum albumin ▲ and vitamin B$_{12}$ •).*

stationary phase ion exchange resins. The functional groups attached to the resins can be described as weak (e.g. WAX, carboxymethyl or diethylaminoethyl) or strong (e.g. SCX, SAX, quaternary ammonium or sulfopropyl). This does not reflect the strength of binding but does reflect their useful operating pH range. Strong ion exchange resins operate over a wider pH range than weak ion exchange resins.

7.10 Size Exclusion Chromatography

Size exclusion chromatography (SEC) is a liquid/liquid partition between the mobile phase and the accessible liquid volume contained within the pores of the stationary phase. There is no physical interaction between the sample and the stationary phase and the separation is also independent of the mobile phase used. Size exclusion can be employed to separate organic and biological polymers (proteins and nucleic acids).

When a sample is applied to a column packed with a size exclusion stationary phase, the molecules will permeate through and around the porous beads in the column. The only parameter that affects the separation is the size and shape of the sample molecules. The pores within the stationary phase have a maximum pore size, which prevents larger molecules from entering into the porous network. The molecules in the sample that are too large to enter the pores of the stationary phase are *excluded* from the stationary phase on the basis of their *size*.

Molecules applied to a size exclusion column will elute from the column with the largest molecules eluting first. These large molecules will only be able to move in the mobile phase around the beads (a smaller volume than the column

164

volume) and they will emerge from the column first. This volume is called the void volume (Vo). Small molecules will also be able to move in the mobile phase around the beads but because of their small size they will be able to enter and permeate into the liquid volume within the beads. Effectively they can move in the total volume of the column (Vt). Depending on their size, other molecules within the sample will be able to access different percentages of the space within the beads. They will elute in a volume (Ve) between the volumes of the large molecules (Vo) and the small molecules (Vt). An effective partition coefficient (Kav) can be calculated using these volumes (see Worked Example 7.4).

Worked Example 7.4: Determining the Kav of a protein in size exclusion chromatography (SEC)

- Blue dextran (Mr 2 000 000), vitamin B$_{12}$ (Mr 1580) and bovine serum albumin were loaded onto a size exclusion column and their elution volumes were measured (see Figure 7.11).
- Blue dextran eluted in 80 ml, vitamin B$_{12}$ eluted in 240 ml and bovine serum albumin eluted in 160 ml.
- The large molecular mass of the Blue Dextran means that it will be totally excluded from the pores of the size exclusion gel eluting in the void volume (Vo). The small molecular mass of the pink component (vitamin B$_{12}$) means that it is able to access the space around the gel (Vo) and the space within the gel to give the total volume of the column (Vt).

$$Kav = \frac{Ve - Vo}{Vt - Vo}$$

Ve is the elution volume of the sample of interest.
Vo is the void volume of the column.
Vt is the total volume column.

$$Kav \text{ (Blue Dextran)} = \frac{80 - 80}{240 - 80}$$
$$= 0$$

$$Kav \text{ (vitamin B}_{12}) = \frac{240 - 80}{240 - 80}$$
$$= 1.0$$

$$Kav \text{ (bovine serum albumin)} = \frac{160 - 80}{280 - 80}$$
$$= 0.5$$

Kav is a number between 0 (Vo) and 1.0 (Vt) without units. It represents a percentage of the stationary gel volume occupied by the sample and in this example bovine serum albumin can occupy 50% of the available pore space in the column because of its relative molecular mass (Mr).

- In SEC the partition coefficient (Kav) of a protein is inversely proportional to the Log_{10} of the proteins relative molecular mass (Mr). Therefore a column can be calibrated measuring the elution volumes of standard proteins of known molecular mass. The Kav can be determined for each protein and a calibration graph constructed plotting Log_{10} Mr on the y-axis against Kav on the x-axis. This calibration graph can then be used to estimate the molecular mass of an unknown protein (also see Chapter 6).

- Size exclusion separates proteins in their native conformation. If a protein is composed of multiple subunits, the total molecular mass will be determined. This is in contrast with denaturing polyacrylamide gel electrophoresis, which estimates the relative molecular mass of a protein's polypeptide chain. In denaturing polyacrylamide gel electrophoresis if a protein is composed of different subunits these will be separated individually (see Chapter 6).

The stationary phase beads in SEC can be made from different inert porous materials for example, dextrans, acrylamide, silica and synthetic polymers. During their construction the pore size can be varied, which results in a wide range of resins that can be used to fractionate molecules of different sizes. Many resins have been strengthened by chemical cross linking, allowing them to be used with high pressure systems.

7.11 Affinity Chromatography

Affinity chromatography is a form of chromatography that is generally applicable to the isolation of components from biological samples. Complex biological polymers such as proteins have their amino acid content derived from their gene sequence. Each protein is synthesized for a biological function and affinity chromatography exploits the biospecificity of a protein. Affinity chromatography was originally designed to isolate a specific enzyme from a sample mixture using the enzyme's substrate bonded to a stationary phase (see Figure 7.12).

Enzymes have evolved specific active sites to bind their substrates with high affinity. In theory, only the enzyme in the applied sample will interact with the substrate bonded to the stationary phase. The other components will percolate through the column without binding. The bound enzyme can then be eluted by adding a high concentration of the substrate (high salt or changing the pH are also used) to compete with and displace the enzyme from the stationary phase. The eluted enzyme can then be collected and analyzed. This form of chromatography represents a powerful method to quickly isolate biological components from a complex mixture. The interactions between specific nucleic acid sequences and proteins, antibodies and their antigens, cell surface receptors and their agonists have also been successfully exploited in affinity chromatography.

Figure 7.12 *Affinity chromatography of an enzyme.*

7.12 Summary

- Chromatography is a versatile technique used to separate components of complex mixtures. It is based upon the distribution (K_D) of compounds between two immiscible phases (i.e. the flowing mobile phase and a stationary phase).
- Chromatography can be used to separate compounds with differences in their polarity, size, volatility and charge using adsorption, size exclusion, affinity, ion exchange or partition chromatography.
- Thin layer chromatography is an inexpensive adsorption chromatography technique for the quick analysis of samples.
- Gas liquid chromatography (GLC) can be used to separate volatile samples
- Octadecylsilane (ODS) is a versatile resin used in reversed phase chromatography. High pressure pumps are required to deliver the mobile phase through columns containing the small diameter stationary phase particles.
- Proteins can be separated using ion exchange, size exclusion and affinity chromatography.

Notes

[1] The stationary phase may be planar (paper chromatography or thin layer chromatography; see Section 7.6) or more often contained within a cylindrical steel, glass or plastic column.

[2] 50% (v/v) methanol is methanol mixed with an equal volume of water that is, 50 ml of methanol mixed with 50 ml of water (see Chapter 1).

[3] A similar technique called HILIC (hydrophilic interaction liquid chromatography) without retention time problems is becoming popular for the separation of polar molecules.

8

CELL CULTURE TECHNIQUES

8.1 Introduction

Bacterial, yeast, plant and animal cells are often grown *in vitro* (i.e. in cell culture medium) to study how they grow, divide or differentiate, how they interact with each other and how they respond to various types of mechanical or pharmacological stimuli. Cell culture provides a controlled environment in which cells can be incubated under defined conditions and treated with growth factors, drugs, toxins or mechanical stimuli in a reproducible fashion. Cultured cells are also widely used in the biotechnology and biopharmaceutical sectors to produce recombinant proteins (e.g. albumin production in bacterial or yeast cell cultures) and antibodies (e.g. in monoclonal hybridoma cell culture technology).

Whatever type of cell is to be cultured, all cell culture methods are based on a common set of principles and requirements:

- The culture should be obtained using a precisely controlled procedure or from a recognized cell culture bank such as the American Cell and Tissue Culture Collection (ATCC: http://www.lgcstandards-atcc.org/).
- Cells need a suitable growth medium containing nutrients to support growth and survival, and this medium must be sterilized before use.
- Cells must be grown under optimized conditions of pH, temperature, humidity, pressure, ionic and osmotic strength.
- Cells must be handled at all times using an appropriate aseptic technique in order to prevent the culture from becoming contaminated by unwanted microorganisms.

In the following sections we will look at basic aspects of aseptic technique applied in culture protocols for the growth of bacterial, animal and plant cells. The main emphasis will be on the handling of bacterial cells, as you are most likely to cover this during the early phase of your studies. As in other areas of laboratory research, accurate dispensing is crucial and general details on pipetting of reagents can be found in Chapter 1. However, this and all other manipulations need to be performed using correct aseptic technique and you are likely to need to use a variety of manual pipetting aids in cell culture work, some

Basic Bioscience Laboratory Techniques: A Pocket Guide, First Edition.
Philip L.R. Bonner and Alan J. Hargreaves.
© 2011 John Wiley & Sons, Ltd. Published 2011 by John Wiley & Sons, Ltd.

Mannual pipette fillers

Turn to draw up liquid

Press to release liquid

Automatic pipette filler with rechargeable battery

Suck and release control buttons

Figure 8.1 Diagrams of pipette aids used in cell culture. The reader is also referred to digital images shown in Section 1.5 (Figure 1.5).

of which are shown diagrammatically in Figure 8.1 (see also Chapter 1). Not only will you need to work using aseptic techniques, the materials you use will also have to be sterile. A selection of the most commonly used methods of sterilization and their application are shown in Table 8.1.

ASEPTIC TECHNIQUE IN MICROBIOLOGY

You must *always* follow local health and safety policy for aseptic technique and for safe disposal of waste.

- Correct aseptic technique will prevent accidental contamination of laboratory cultures from external sources (e.g. hands, clothes and environment) and prevent contamination of you and others working in the laboratory.
- Always wear an appropriate lab coat and wash your hands before entering and leaving the microbiology laboratory.
- Only use sterilized equipment and media and always decontaminate any used items (e.g. in bleach) before discarding them.
- Glass pipettes will be plugged with cotton wool and sterilized in a metal tin by autoclaving. Flame the opening of the tin before removing a pipette to avoid contamination of the remaining pipettes.

170

Table 8.1 Common methods of sterilization of materials and equipment used in cell culture

Method	Comments	Application
Heat treatment	Red heat sterilization. This involves heating for a few seconds in a Bunsen flame. Simple and effective.	Sterilization of metal implements such as loops, forceps, needles, and so on. Glass rods and spreaders are dipped first in ethanol, which is then burnt off by brief exposure to the flame.
	Baking for 2 h or more at 160 °C or higher (dry heat sterilization). Time consuming.	Sterilization of glass ware
	Autoclaving. This involves heating to about 120 °C at high pressure for 20 min. Method of choice for routine sterilization of most materials	Media, glassware, pipette tips and so on. NB: Cannot be used for heat sensitive media or plastics.
Chemical treatment	The main example is the use of disinfectants	Used to treat spillages and to disinfect used materials. They need at least 10 min to sterilize contaminated areas
Filtration	Sterile filters (pore size 0.2 or 0.45 μm) can be assembled in the filter unit and sterilized by autoclaving. They can also be purchased as a one-use pre-sterilized disposable unit.	This is the preferred method of sterilization for heat sensitive media and reagents.
Radiation	Materials can be sterilized by ultraviolet or ionizing radiation (e.g. gamma rays) but the latter can only be done on an industrial scale.	This method is used to treat most disposable cell culture materials (culture flasks, pipettes, pipette tips, etc.). Materials are discarded after a single use.

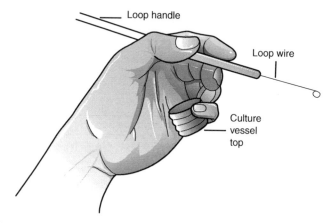

Figure 8.2 *Holding the loop and culture vessel lid during manipulations. The lid of the culture vessel is carefully unscrewed using the little finger, while holding the loop between the thumb and first finger as shown. Always follow local rules on the use of personal protective equipment (e.g. gloves, glasses, etc.)*

- Never place the culture vessel lid (metal, foil or cotton wool plug) on the bench surface as it will get contaminated or contaminate the bench area. You should remove and hold the vessel lid with the little finger of one hand during manipulations, which becomes easy with practice (Figure 8.2).
- When working with liquid cultures, pass the open mouth of the glass vessel quickly through a Bunsen flame after removing and before replacing the lid. Working close to the up draught of a Bunsen flame reduces the risk of entry of air contaminants into the vessel (Figure 8.3).
- Agar-based solid media are prepared in laminar flow cabinets (Figure 8.4) which enables solid media, which are heated to form molten agar, to solidify by cooling for prolonged periods with the Petri dish lid removed, without risk of bacterial contamination from the air.
- While bacterial cultures can be handled on the bench,[1] a laminar flow cabinet should be used for all manipulations with plant and animal cell cultures.
- Before starting laboratory work, make sure that your bench area is clear of all obstacles, wipe it down with an appropriate disinfection agent (e.g. 70% v/v ethanol) and place all items in convenient locations with a clear workspace in front of you.

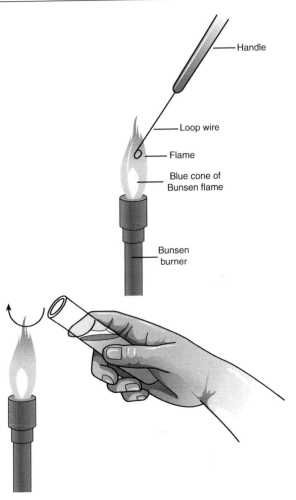

Figure 8.3 *Flaming the loop and the culture vessel opening. The loop wire should be positioned at an angle to the hottest part of the flame, which is just outside the blue cone. When the loop wire is red hot, remove from the flame and allow it to cool for a few seconds before collecting cells. To reduce the risk of contaminants entering the culture vessel, after removing the lid (as shown in Figure 8.2) pass the neck of the vessel briefly through the flame before carrying out each manipulation.*

Figure 8.4 *Schematic illustration of the controlled air flow in a laminar flow cabinet. This type of set up is used for preparation of many media for bacterial culture and for most cell culture manipulations with plant and animal cells.*

- Any spillage on the bench should be covered with disinfectant for at least 10 min to kill all organisms before mopping up.
- While unlikely to be working with hazardous strains of bacteria at this stage in your career, it is important to be aware of their classification according to the hazard they represent to human health (Table 8.2).

8.2 Growth and Maintenance of Cells in Culture

The main focus of this section will be on techniques applied to bacterial culture. However, its applications in eukaryotic cell culture will also be briefly discussed.

Table 8.2 *Classification of micro organisms into hazard group*

Hazard group	Likelihood of causing human disease	Groups at risk
1	Unlikely to cause disease.	—
2	May cause disease.	Laboratory workers. Low hazard to the community.
3	May cause severe disease.	May be a serious hazard to laboratory workers and may spread to the community.
4	Causes severe disease.	Serious hazard to laboratory workers and potentially high risk to the community

As you are no doubt aware, not all bacteria are pathogenic. There are numerous bacteria on exposed surfaces of the body (skin, hair, gut, etc.) that are beneficial to health and prevent invasion by pathogenic bacteria. Indeed, this is a selling point of certain health foods (e.g. probiotic foods and yoghurts). Likewise, there are beneficial and pathogenic strains of bacteria in the environment (e.g. in soil and sewage).

One way to detect beneficial and pathogenic bacteria is to grow samples in bacterial growth media (singular '*medium*'), which may be solid or liquid. Media may be non-selective, selective and/or differential. Non-selective media allow the growth of a range of bacteria, whereas selective media only allow the growth of specific bacteria or groups of bacteria. Differential media can be used to identify a specific sub group of bacteria, for example due to a colour difference in growing colonies. Examples of typical growth media and their applications are shown in Table 8.3.

8.2.1   Solidified Media

Many bacteria can be applied to and cultured on the surface of agar-based media. An advantage of this technique is that each single viable cell inoculated onto the agar surface can divide to form a colony of identical cells that is visible to the naked eye. This technique forms the basis of most bacterial isolation and purification methods, including procedures such as streak dilution, spread plate and pour plate, which are discussed in more detail below. The main component is normally agar, which is melted at high temperature after which other

Table 8.3 Selected examples of typical growth media for cell culture

Medium	Cell type	Composition[a]
Dulbecco's modified Eagle's medium (DMEM)	Animal cells (mammalian) grown as monolayers or in suspension	Inorganic salts, amino acids, glucose, sodium pyruvate, phenol red pH indicator.
Murashige and Skoog medium	Plant cell cultures (various types)	(a) Macronutrients (to provide the 6 major elements: phosphorous, nitrogen, potassium, calcium, magnesium and sulfur);
		(b) micronutrients (Iron, manganese, zinc, boron, copper and molybdenum);
		(c) vitamins (e.g. thiamine, nicotinic acid, pyridoxine)
		(d) amino acids or other nitrogen supplements (e.g. glycine, glutamine, asparagine, arginine, cysteine or tyrosine),
		(e) growth regulators (e.g. auxins, cytokinins, gibberellins or abscisic acid);
		(f) sugars (e.g. sucrose or other sugars);
		(g) Undefined organic components (e.g. coconut milk, casein hydrolysate)

| Selective enrichment medium | Extreme halophiles (bacteria that grow in extreme salt environments – halobacteria) | High NaCl (about 250 g/L), casaminoacids (a source of S, N and P), yeast extract (as a source of growth factors), tri-sodium citrate (as a C and energy source) and various inorganic salts (as a source of K^+, Mg^+, Fe^{2+} and S) |
| Minimal medium for *Bacillus megaterum*. | An example of a basic medium for the culture of a heterotrophic bacterium (i.e. a bacterium that requires an exogenous supply of organic compounds in its growth medium in order to survive) | (a) Sucrose (about 1% w/v; as a C and energy source); (b) various phosphates of potassium and ammonium (as buffer components and a source of K^+, N and P); (c) various inorganic salts (as a source of Mg^{2+}, Mn^{2+}, Fe^{2+} and S) |

Shown are examples of growth media for the culture of animal, plant and bacterial cells with a brief overview of their composition.
[a]Components may vary (e.g. serum supplements in many animal cell cultures, agar in solid media formulations, blood in 'blood agar' plates, etc.).

nutrients are added and the agar is allowed to solidify by cooling it to room temperature.

8.2.1.1 *Streak Dilution*

This is one of the primary techniques used for obtaining pure microbial cultures. It involves spreading bacteria collected on a metal loop over an agar surface until individual cells fall off the loop in some areas. In theory, each individual cell can divide and give rise to a colony of identical cells. However, as the source material (typically exponentially growing cells in liquid medium) usually contains a very high cell density, a series of steps is required to dilute the applied sample.

STREAKING DILUTION

As most liquid sources of bacteria will contain between 10^5 to 10^7 bacteria per ml a loop full of bacteria could contain 10^3 to 10^5 cells, which would form enough colonies to completely cover a 10 cm diameter agar plate. Further steps that are needed to obtain colonies from individual cells are summarized below.

- Sterilize the loop in a Bunsen burner flame for a few seconds (Figure 8.3).
- While the loop cools, pick up the culture tube or flask (containing exponentially growing cells) with the other hand and remove the cap or plug carefully using the little finger of the hand holding the loop (Figure 8.2).
- Place the loop into the liquid medium (broth) to collect the cells for transfer.
- Replace the lid on the culture vessel and set it down before lifting the agar plate with the non-dominant hand so that the lid faces upwards (Figure 8.5).
- Remove the Petri dish lid using the non-dominant hand (Figure 8.5a).
- Streak the plate (Figure 8.5b) then put the lid down before resterilizing the loop.
- Reorientate the Petri dish at 45–90 degrees to the original position.
- Restreak from the top end of the first streak so that there is a streak dilution of the original streak applied (Figure 8.5b).
- Replace the lid again and resterilize the loop.
- Lift the lid again and restreak as before at an angle to the second streak (Figure 8.5c).
  Close the lid and resterilize the loop before setting it down.

Several streak dilutions can be performed in this way (Figure 8.5b to e) before the streaked plate is incubated under appropriate conditions until colonies appear (Figure 8.5f).

NB: Always remember to label Petri dishes clearly with your name, the name of the bacterial strain, and so on, which should be written on the bottom plate so as not to obscure visual examination of the colonies.

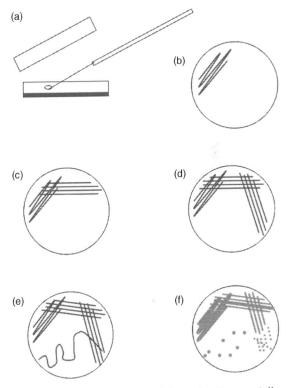

Figure 8.5 *Streaking an agar plate with bacteria. Exponentially growing cells are collected in a loop and streaked onto an agar plate as directed in the Practical hints and tips – streaking dilution information box (a to e). After a suitable incubation period, many colonies appear, some of which are individual (f). Always follow local rules on the use of personal protective equipment (e.g. gloves, glasses, etc.)*

8.2.1.2   *Spread Plate Dilution*

In this method a small volume (50–500 µl) of serial dilutions (see Chapter 1) of cells in liquid culture medium or an appropriate sterile diluent, is applied to the surface of solid agar medium in a Petri dish. It is then evenly spread over the whole agar surface using an L-shaped glass spreader. Prior to this the spreader is sterilized by dipping it into a beaker containing 70% (v/v) ethanol, the excess ethanol is allowed to drain and the remainder evaporated off by gently passing the rod through a Bunsen flame. Once cooled, it can then be used to spread the cell suspension, as indicated in Figure 8.6. The Petri dish lid should be removed for the minimum time possible and breathing onto the plate should obviously be avoided.

This is one of the main methods used to quantify the number of viable cells or colony-forming units (CFUs) in a bacterial culture sample. It is typically done using a range of dilutions until an easily measurable colony density of between approximately 20 and 200 per plate is obtained. If this requires dilution, multiplication by the dilution factor will give the actual viable cell count as CFUs per ml of original culture (see Worked Example 8.1).

Worked Example 8.1: Using the spread plate technique to determine viable cell counts

A volume of 0.1 ml of serial 10-fold dilutions of a bacterial culture were applied to Petri dish culture plates and incubated at 37 °C for 24 h. The neat and 1/10 dilutions had too many colonies to count. The 1/100 dilution produced 165 colonies. The 1/1000 dilution produced only 15 colonies and no colonies appeared at further dilutions. Estimate the number of viable cells per ml of the original culture medium.

- Assuming that each colony arose from a single viable cell, take the colony count for the 1/100 dilution and multiply by the dilution factor to obtain the number of viable cells in 0.1 ml of culture.

$$\text{CFUs in } 0.1 \text{ ml} = 165 \times 100 = 16\,500$$

Multiply by 10 to obtain the viable cell number per ml

$$\text{CFU/ml} = 16\,500 \times 10 = 165\,000$$

Applications:
- This type of approach can be used to estimate the viable cell counts for different types of cultures obtained from clinical isolates, hospital equipment, environmental samples (soil, air and water), and so on.

Spread dilution

Pipette

1. Apply a small volume of exponentially growing cells in liquid medium

Lid

Base

2. Flame the spreader previously dipped in 70% v/v ethanol

3. Remove lid and quickly spread cell suspension evenly on agar plate. Rotate spreader until medium is absorbed by agar

4. Replace lid and incubate under appropriate conditions

Petri dish lid

Petri dish base

Colonies should be evenly spread across the agar medium surface

Figure 8.6 Spread plate dilution of bacterial cultures. Exponentially growing cells are applied by pipette onto the surface of the solid agar medium and spread using a glass spreader as indicated in the figure itself. Always follow local rules on the use of personal protective equipment (e.g. gloves, glasses, etc.)

- If performed using selective media it can aid in the identification process.
- It can also be used to study the ability of antibiotic molecules to inhibit colony formation.

8.2.1.3 Pour Plate Dilution

In this procedure a known volume of cells grown in liquid medium are mixed with molten agar medium, usually in screw-cap bottles containing a volume sufficient to pour one agar plate (15–20 ml). The agar will only remain molten at temperatures above about 45 °C and, although most bacteria are not killed as a result of a few minutes exposure at 45–50 °C the procedure can be damaging to bacteria taken from low temperature environments. The agar is maintained in its molten form in a water bath. The cap is removed and culture medium added using the aseptic techniques discussed earlier. The cap is immediately replaced and the cells carefully mixed with the molten medium, which is then quickly poured into the Petri dish and allowed to solidify by cooling with the lid on. The plates are then incubated under appropriate conditions to allow colony formation and counting.

POUR PLATE DILUTION

- Cells should be mixed with the molten agar medium by rotating the bottle between the palms of the hands, as shaking can cause frothing of the medium.
- Colonies which are much smaller than those observed on a streaking dilution plate will form within the solid medium rather than on its surface, giving the advantage that a much larger number of colonies can be counted than with streaking dilution.
- As these colonies are also different in shape to those observed on a streaking dilution plate, it is often the case that an additional thin layer of molten agarose is placed on top of the previously poured medium to prevent the growth of wider colonies on the medium surface.
- Remember to disinfect or sterilize the glass rod before setting it down.

8.2.2 Other Applications of Solid Media Culture Methods

8.2.2.1 Bacteriophage Detection

Bacteriophages ('phages') are viruses that infect bacteria. Bacteriophage-mediated lysis of bacteria grown on agar plates is often used in laboratory classes to

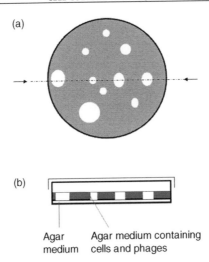

(a)

(b)

Agar
medium

Agar medium containing
cells and phages

Figure 8.7 *Bacteriophage detection on agar plates. Shown is a schematic diagram of an aerial (a) and cross sectional view (b) showing the appearance of clear plaques where active phages are present in a bacterial culture on solid medium.*

demonstrate the detection of viruses and the determination of their quantity in an extract or isolate. Although phages are too small to be seen using a light microscope, their presence in a bacterial culture can be detected by measuring their ability to cause lysis (destruction) of bacteria grown on agar plates.

For this reason, bacterial cells cultured in liquid medium are dispersed in molten agar with a known number of bacterial phages, as described above. This is then poured out to form a thin layer on the surface of a layer of the same agar in a Petri dish. If no phages are present, a continuous 'lawn' of bacteria grows throughout the top agar layer following incubation under appropriate conditions.

When virulent phages are present, clear circular areas termed 'plaques' are observed in areas where a phage is present (Figure 8.7). Each plaque is the result of a single infective phage. Virulent phages are capable of infecting a cell, replicating within the cell and causing cell lysis within an hour. On lysis, more phages are released to infect neighbouring cells growing in the surrounding agar medium. As the lytic cycle of virulent phages is often very short, plaques need to be counted within a short time window as the plaques get larger over time and may begin to fuse with each other making accurate counts difficult to achieve.

Counting can be facilitated by viewing the plate against a dark background and marking the plaques on the base of the plate with a marker pen. The number of plaque forming units (PFUs) per ml of virus culture can be determined in a manner similar to that described earlier for bacterial colony forming units, taking into account the volume and dilution (if any) of viral extract at the point of application to the molten agarose medium (Figure 8.7).

8.2.2.2 Plant Tissue Culture

Plant tissue can also be cultured on agar by removing small areas of tissue from a plant (explants), sterilizing their surface (e.g. by rinsing for a few minutes in a solution of 10 % [w/v] sodium hypochlorite), after which it is transferred to an agar-plate containing a suitable growth medium to support the growth of a mass of undifferentiated cells or 'callus' (Figure 8.8). Using an appropriate combination of growth medium and plant hormones it is possible to stimulate the growth of new plantlets. Such approaches can be used to determine the requirements for the regulation of plant growth and development in experimental work. They can also be used to produce infection free stocks of rare, endangered or commercially important plants to allow their reintroduction to the environment at a later date. Plant cells derived from callus tissue can also be transferred to liquid culture (see below).

8.2.3 Liquid Media

In situations where the rate of growth is to be determined (e.g. by constructing a growth curve) or there is a requirement to generate large cell numbers (e.g. in the large scale production of recombinant proteins), cells can be grown in a suitable liquid medium usually with agitation. Again, if you come across this technique in classes you are most likely to be handling bacterial cell cultures and hence the main emphasis in this section will be on microbiological examples, but with some discussion of liquid culture of eukaryotic cells.

8.2.3.1 Cell Suspension Cultures

In microbiological applications, bacterial cells are typically grown in sterile conical flasks in a variety of sizes (200 ml to 2 litres) depending on the cell numbers required. Flasks are rotated on an orbital shaker operating at the required speed (usually 20–250 rpm) and temperature, which depend on the properties of the culture being grown. Larger vessels are often referred to as fermenters or bioreactors in the field of biotechnology.

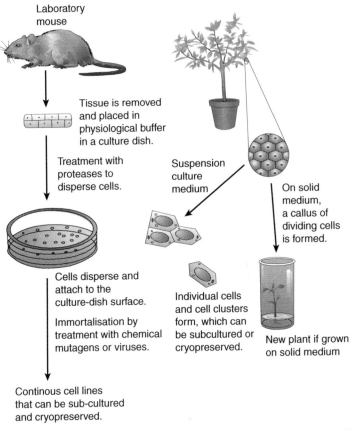

Figure 8.8 *Typical work flow for the generation of plant and animal cell cultures. Note that certain animal cell types (eg. white blood cells) are grown in suspension culture.*

If the organism being cultured is an aerobic bacterium, it is important to limit the medium to a maximum of 1/5th of the flask volume in order to provide adequate surface area for absorption of air. Larger scale cultures may require more vigorous stirring (e.g. using a sterilized magnetic stirrer) and the

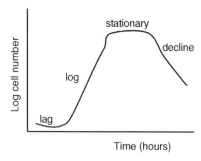

Figure 8.9 *Typical growth curve for cultured cells. All cell cultures grow at their own characteristic rates but usually they conform to a specific pattern of growth involving a growth curve with a lag period followed by rapid growth, stationary and decline phases. Adapted from Alberts, Molecular Biology of the Cell 5th Edition, Garland Press.*

introduction of sterile air through a filter system. The most simple air filter is a sterile glass wool or cotton wool plug placed in the neck of the flask. Alternatively a commercially available filter unit can be used, which would normally have a pore size of 0.2 μm. However, it is also important to note that cultures of anaerobic bacteria are normally grown without agitation so as to reduce aeration of the cells.

Batch culture is the most common liquid cell culture method and is widely used in small-scale laboratory work. Using aseptic technique, cells are inoculated into a vessel containing appropriate growth medium after which they increase in number following a predictable pattern with four main phases, referred to as lag (slow growth), log (exponential growth), stationary (no net growth) and decline (death), as indicated in Figure 8.9.

Cell growth can be measured in a number of ways including counts in a haemocytometer chamber (see Chapter 1), in an electronic particle counter (e.g. a Coulter® counter), by taking turbidity measurements of the cell suspension in a spectrophotometer (i.e. optical density or attenuance of light; see Chapter 3), and so on, as outlined in Table 8.4. For any cell culture it is possible to estimate the growth rate constant or generation time according to the equations below (see Worked Example 8.2):

$$\text{Growth rate } (\mu) = \frac{2.303 \, (\log N_2 - \log N_1)}{t_2 - t_1}$$

N_1 and N_2 represent cell counts determined at an early (t_1) and late (t_2) time point on the exponential part of the curve. If times are measured in hours the units of μ would be expressed 'per hour'.

Table 8.4 *Methods of measuring cell growth*

Methods	Principles	Disadvantages
Microscopy counts	A cell count is made on a diluted cell suspension in a haemocytometer chamber (see Chapter 2)	Time consuming if multiple counts are required. Living and dead cells may be difficult to distinguish. Clumping of cells may affect accuracy. Viable cell counts may be possible using non cell permeable DNA binding dyes (e.g. SYTOX green and propidium iodide) to detect damaged cells. However, an instrument with fluorescence detector is required
Electronic particle counting	Coulter counters detect changes in electrical resistance as cells pass through a channel, allowing very rapid measurements of total cell counts.	It may be difficult to distinguish living, and dead cells, cell clumps and other particles. A sophisticated alternative is to use a flow cytometer which can sort particles as well as count them.
Cell culture based methods	Cell cultures work on the assumption that under appropriate conditions an individual viable (living) cell can multiply to form a colony on agar plates or a measurable change in turbidity (cloudiness) in liquid culture. The main advantage of this approach is that only viable cells are counted.	The true viable count will be underestimated if not all cells are able to grow under the conditions used. Cell clumping and dilution errors can lead to further inaccuracies. Sterile media, equipment are needed. Long incubation times of up to 3 d may be required to produce measurable growth.

The time it takes for a cycle of cell replication referred to as the doubling time (T2) or generation time (g) is given by the equation below:

$$T2\ (g) = \frac{0.301(t_2 - t_1)}{\log N_2 - \log N_1}$$

Worked Example 8.2: Growth rate and doubling time

If you counted 4000 cells at the first time point and 128 000 cells after 2 hours the growth rate is calculated from the equation above as:

$$\mu = \frac{2.303 \times (\log 128\,000 - \log 4000)}{2}$$

$$= \frac{2.303 \times (5.11 - 3.60)}{2}$$

$$= \frac{2.303 \times 1.51}{2}$$

$$= 1.74\ \textbf{per hour}$$

This indicates the number of times per hour that the population will double during exponential growth

The *doubling time* would be calculated as follows:

$$g = (0.301 \times 2) \div (\log 128\,000 - \log 4000)$$

$$= \frac{0.602}{1.51}$$

$$= 0.4\ \textbf{hours}\ (24\ \text{minutes})$$

Therefore the population doubles every 24 minutes during the exponential growth phase.

8.2.3.2 *Alternatives to Batch Culture*

Continuous culture can be used when cells need to be maintained in the exponential growth phase for long periods of time. It involves continuously removing a fixed volume of culture and replacing it with new culture medium so that cells are diluted and the remaining cells have sufficient nutrients to maintain exponential growth. Any increase in cell number due to proliferation must be offset by a similar loss of cells due to the dilution with fresh medium. While it is useful to be aware of this approach it is unlikely that you will be exposed to this technique in the early stages of your studies.

Continuous cultures are more difficult to maintain and are more prone to contamination than batch cultures, as they need rigorous aseptic technique and the main culture vessel is connected to at least two additional vessels, one to introduce the culture medium and the other to collect the culture that is removed. However, it has the advantage that a large number of cells at the same stage of growth can be maintained over long time periods, making them useful for physiological and biochemical studies. The process can be automated (computer controlled) and is often used to determine conditions for continuous culture in larger scale bioreactors used in the biotechnology industry (e.g. for the production of antibodies or recombinant proteins).

8.2.4 Liquid Culture of Eukaryotic Cells

8.2.4.1 *Suspension Culture*

Cells from a range of plants and animals have been successfully grown in suspension culture. Procedures are similar to those used for handling bacterial cells except that aseptic manipulations for small-scale cultures are normally carried out using a laminar flow cabinet (Figure 8.4), which provides an aseptic environment using a filtered air circulation system designed to minimize the risk of microbial contamination.

A wide range of plant cell cultures has been established from different plants and plant tissues. Typically, a fragment of callus tissue (obtained as described in Section 8.2.3) is transferred under aseptic conditions to liquid growth medium in a conical flask, which is then placed in an orbital shaker set at a speed that encourages cell dispersal to form a population containing a mixture of individual cells and cell clusters. A homogeneous culture of individual cells is unlikely to be obtained due to the formation of plasmodesmata between many plant cells.

Certain animal cells such as white blood cells are also grown in suspension culture though the majority of animal cell cultures involve cells grown in the form of a monolayer attached to a plastic culture dish surface (e.g. fibroblasts, epithelial and endothelial cells, neurons, etc.).

8.2.4.2 *Monolayer Culture of Mammalian Cells*

Most animal cell types are cultured using sterile plastic ware, in which the internal plastic surface of the culture dish or 'flask' is usually treated chemically or by microwave irradiation to produce a surface that increases cell attachment to it. If the cells are left to incubate for too long the monolayer becomes confluent

Table 8.5 Selected examples of animal cell cultures

Cell line	Source	Applications
Hybridoma	Fusion of a lymphocyte with a myeloma (a type of tumour) cell.	Production of monoclonal antibodies.
Fibroblasts	Connective tissue or skin	Studies of fibroblast function in tissue organization and wound healing. These are normally finite cell lines (i.e. they survive 40–50 divisions) but transformed (immortalized) cell lines are also available.
Keratinocytes	Epidermis (The major cell type in the outer layer of the skin)	Useful for studies of skin cell function, wound healing, and for the creation of artificial skin. These may exist as finite or immortalized cell lines.
Lymphocytes	White cells from the blood	These can be isolated freshly from but used immediately to study T cell functions. Transformed (tumour) cell lines are also available.
Hepatocytes	Liver (The major cell type)	Useful for studies of liver cell function. Primary cultures are short-lived but established (immortalized) cell lines also exist.

| Neuroblastoma | From neuronal tumours in the central nervous system. | Primary cultures of neurons are short-lived. Established (tumour) cell lines such as neuroblastoma divide indefinitely and can be used in studies of neuronal cell function. They can be induced to differentiate into a mature neuronal phenotype by appropriate manipulation of the growth medium – useful in studies of development. |
| Glioma | From glial tumours in the central nervous system. | Primary cultures of glial cells can be obtained from neural tissue but these are normally short lived. Glioma cell lines are immortal and can be used in studies of glial cell function and differentiation (development). |

Shown is a selection of the types of cell that can be grown in culture. Many cell types are available commercially. These are kept in cryopreservation vessels and dispatched to investigators on dry ice.

(i.e. the cells cover the whole culture dish surface) and cells can become detached and/or lose viability as nutrients in the medium become limiting. To avoid this, they are usually 'fed' every couple of days, which involves replacing some of the spent medium with fresh medium. Before becoming confluent the monolayer is detached from the culture dish surface by brief incubation with trypsin (see Section 5.2.1), which degrades proteins that enhance cell attachment to the culture dish surface. The cells are gently removed by pipette to form a suspension, a fraction of which (usually between a tenth and a third, depending on how rapidly the cells divide) is diluted in fresh growth medium. This process of sub culture is often referred to as a '*passage*'.

In some studies the cell culture dish needs to be coated with a protein found in the extracellular matrix (e.g. collagen, fibronectin or laminin) in order to simulate *in vivo* environments more closely. As an additional measure against contamination, antibiotics such as penicillin and strepto-mycin are often added as supplements to culture media. As for bacterial cell culture, all equipment and materials are sterilized prior to their use in eukaryotic cell culture.

Various types of mammalian cell cultures are used in research and in the biotechnology industry. Important features about the main types of mammalian cultures and their production are summarized in Table 8.5 and Figure 8.7.

8.2.5 Cryopreservation of Cells

A major problem with cultured cells is that, if they are sub-cultured repeatedly over long periods of time, genetic drift can occur. This may lead to changes in cell phenotype and physiology. It is therefore necessary to cryopreserve stocks of cells after low numbers of subculture, in order to preserve the original features of the cell line. The most effective way of doing this is to gradually freeze exponentially growing cells in growth medium containing cryopreservation agents such as dimethyl sulfoxide (DMSO) at 5–10 % (v/v). Cells frozen in this way can be stored indefinitely at $-196\,°C$ to $-150\,°C$ in liquid nitrogen or in its vapour phase. When needed, cryopreserved cells are rapidly thawed and returned to normal growth conditions.

8.3 Summary

- Cells can be isolated and cultured from a range of sources in liquid or on solid media.
- In culture, cells can be used to study a range of physiological phenomena and the effects of toxins or therapeutic agents.

- They can be used on a larger scale in the production of recombinant proteins and/or drugs.
- It is essential that good aseptic technique is used when handling cultured cells and that stocks are cryopreserved at an early stage so as to avoid genetic drift.

Notes

[1] As bacterial cultures grow very rapidly, they are less likely to become contaminated by other bacteria than animal and plant cultures, which are much slower growing.

SUGGESTIONS FOR FURTHER READING

Most biochemistry and cell biology textbooks will contain some supplementary information on selected laboratory techniques. For example:

- Berg, J.M., Tymoczko, J.L. and Stryer, L. (2006) Biochemistry (6th edn) ISBN: 071676766X WH Freeman.
- Nelson, D.L. and Cox, M.M. (2009) Lehninger Principles of Biochemistry (5th edn) ISBN: 0-7167-7108-X WH Freeman.
- Alberts, B., Johnson, A., Lewis, J., Raff, M., Roberts, K., and Walter, P. (2007) Molecular Biology of the Cell (5th edn) ISBN: 978-0-8153-4105-5 Garland Press.
- Lodish, H., Berk, A., Kaiser, C.A., Krieger, M., Scott, M.P., Bretscher, A., Ploegh, H. and Matsudaira, P. (2008) Molecular biology of the Cell ISBN: 0-7167-7601-4 WH Freeman.

More detailed information on general laboratory techniques can be found in:
- Walker, K. and Wilson, J. (2000) Principles and techniques of practical biochemistry (5th edn) ISBN: 052165873X Cambridge University Press.

Supplementary reading for specific chapters

Chapter 1:
Beynon, R.J. and Easterby, J.S. (2003) Buffer solutions ISBN: 0199634424 Biosis Sci Publishing Ltd., Oxford, UK

Chapter 2:
Slayter, E.M. and Slayter, H.S. (1997) Light and electron microscopy ISBN: 0521339480 Cambridge University Press, Cambridge, UK.

Chapter 3:
Gore, M. (2000) Spectrophotometry and Spectrofluorimetry: A Practical Approach ISBN: 0199638128 Oxford University Press, UK.

Basic Bioscience Laboratory Techniques: A Pocket Guide, First Edition.
Philip L.R. Bonner and Alan J. Hargreaves.
© 2011 John Wiley & Sons, Ltd. Published 2011 by John Wiley & Sons, Ltd.

Chapter 4:

Lindley, D.V. and Scott, W.F. (1984) New Cambridge Statistical Tables. 2nd edition. ISBN: 978-0521484855, Cambridge University Press, Cambridge, UK.

Quinn, G.P. and Keough, M.J. (2002) Experimental Design and Data Analysis for Biologists. Cambridge University Press, Cambridge, UK.

Dytham, C. (2003) Choosing and Using Statistics. A biologist's guide. 2nd edition. Blackwell Science Publishing, Oxford, UK.

Doerge, R.W. and Bremer, M. (2009) Statistics at the Bench: A step by step handbook ISBN: 978-0879698577 Cold Spring Harbor Laboratory Press, Cold Spring Harbor, New York.

Chapter 5:

Graham, J. (2001) Biological centrifugation ISBN: 1859960375 Garland Press.

Chapter 6:

Hames, B.D. (2002) Gel Electrophoresis of Proteins: A practical approach. 3rd edition ISBN: 019963640 Oxford University Press, Oxford, UK.

Brown, T.A. (2000) Essential Molecular Biology Vol 1 (2nd edn) Practical approach ISBN: 0199636427 Oxford University Press, Oxford, UK.

Chapter 7:

Poole, C. F. (2002) The essence of chromatography ISBN: 0444501991 Elsevier Press.

Bonner, P.L.R. (2007) Protein purification (the basics) ISBN: 0415385113 Taylor and Francis.

Chapter 8:

Harrison, M.A. and Rae, I.A. (2001) General Techniques of Cell Culture. Handbooks in Practical Animal Cell Biology ISBN 978-0521574969 Cambridge University Press, Cambridge, UK.

Rhodes, P.M. and Stanbury, P.F. (2001) Applied Microbial Physiology: A Practical Approach ISBN: 0199635781 Oxford University Press, Oxford, UK.

Detailed information on laboratory techniques can be usually found on the websites of the manufacturers of laboratory equipment. Examples include:

- http://beckman-coulter.com for information about centrifugation
- http://www.phenomenex.com for information on chromatography columns

- http://www.gehealthcare.com for information on protein purification systems and other bioscience equipment
- http://www.bio-rad.com for information on electrophoresis and other bioscience equipment
- http://www.waters.com for information on HPLC and other chromatography systems
- www.sigmaaldrich.com for reagents and cell culture protocols.

More information on statistics and statistical tables can be found at:

- http://www.dur.ac.uk/stat.web/
- http://www.statsoft.com/textbook/distribution-tables/#z
- http://web.abo.fi/fak/mnf/mate/kurser/statistik1/StaTable.pdf

INDEX

Basic Bioscience Laboratory Techniques: A Pocket Guide, First Edition.
Philip L.R. Bonner and Alan J. Hargreaves.
© 2011 John Wiley & Sons, Ltd. Published 2011 by John Wiley & Sons, Ltd.

Printed and bound by CPI Group (UK) Ltd, Croydon, CR0 4YY